PROF. ARTHUR B. BOWEN
1926 - 1987
MEMORIAL COLLECTION

THE CONCEPT OF HEAT AND ITS WORKINGS SIMPLY EXPLAINED

by Morton Mott-Smith, Ph.D.

Dover Publications, Inc., New York

This Dover edition, first published in 1962, is an
unabridged and unaltered republication of the work
originally published by D. Appleton and Company
in 1933 under the title *Heat and Its Workings*.

International Standard Book Number:0-486-20978-4

Manufactured in the United States of America
Dover Publications, Inc.
180 Varick Street
New York, N. Y. 10014

INTRODUCTION

WHAT are these calories? Can they be taken in tablet form, or do we have to mix them with our food? Neither! Calories are heat units. They come already mixed with the food. They constitute its *energy* content. We cannot live without them. Although too many of them may not be good for those who wish to remain slender, or even for those who wish to remain healthy, a sufficient number of them we must have, for without them we die. Every animal, every plant, everything that lives is looking for calories. But not only the living world, the non-living must also have its quota. Everything that happens is due to the flow and transformation of energy. Without its calories in the right places, and in the right numbers, the universe itself dies.

The gods did well to bind Prometheus to the rocks when he stole fire from heaven, for by this act he enabled men themselves to become like gods and to rule the elements. Control fire, and you control everything. The discovery of fire was indeed the greatest achievement of primitive man, the one thing that more than any other lifted him from the level of the beast, and gave him dominion over the earth. Heat is his most powerful and obedient servant. It cooks his food, warms his dwelling, moves his machinery, and performs a myriad of useful services for him.

Heat has a profound effect upon the states and conditions of matter. Every property of matter is altered by a change in temperature—the density, pressure, hardness, strength, electrical conductivity, in short everything that one can mention. Whatever happens in this physical world of ours, one has always to reckon with the temperature or the quantity of heat involved.

Some of these effects of heat, and some of its ways that can be helpful to man, are described in the following pages.

Acknowledgments are due to Mr. James T. Barkelew, who painstakingly read the manuscript of this book, corrected some errors, and made many valuable suggestions; and to Mr. Carl Nagashima, whose skillful pen lettered the diagrams.

M. M.-S.

CONTENTS

ILLUSTRATIONS

THE CONCEPT OF
HEAT AND ITS
WORKINGS
SIMPLY
EXPLAINED

I
The Temperature Sense

There is a very old and simple experiment described by John Locke in 1690 in his *Essay Concerning Human Understanding*, and it was doubtless old even then.

Let the two hands be immersed, one in a basin of hot water, the other in a basin of cold water. Then after a time plunge both hands into a bath of intermediate temperature. The water will seem cold to the one hand and hot to the other. What is the explanation? Says Locke:[1] "If the sensation of heat and cold be nothing but the increase or diminution of the motion of the minute parts of our bodies, caused by the corpuscles of any other body, it is easy to be understood, that if that motion be greater in one hand than in the other, if a body be applied to the two hands, which has in its minute particles a greater motion than in those of one of the hands, and a less than in those of the other, it will increase the motion of the one hand and lessen it in the other, and so cause the different sensations of heat and cold that depend thereon."

Even at that early date Locke conceived of heat as a motion of the minutest particles of bodies. In this he was much ahead of his times, for this view was not definitely adopted by science until a hundred and fifty years later. Yet his explanation could not have been more perfect had it been written yesterday.

According to this theory then, a body seems hot when heat is conveyed to the hand, and cold when heat is conveyed away from the hand. The temperature of the human body thus forms the dividing line between what we call hot, and what we call cold, and the position of this line is variable. The distinction cannot therefore be an objective one. Outside of our

[1] Book II, Chapter VIII, § 21.

subjective impressions, there are no cold bodies.   All bodies are
hot, and their degrees of hotness are measured up from some
absolute zero of complete heat absence, which lies far below the
temperature of the human body.

Another old experiment is this: If we touch successively
hotter and hotter bodies, the sensation of heat increases for a
time, but soon becomes painful.   When the bodies are very
hot, we cannot distinguish their temperature differences at all,
but receive only the stinging sensation of a burn.   Similarly if
we touch successively colder bodies, the sensation of cold
increases up to a certain point, and then becomes painful.   An
extremely cold body produces a burn just like a hot one, and to
the touch is indistinguishable from it.   This fact was mentioned
by Francis Bacon in his *Novum Organum* in 1620.

In sensation, then, there is a maximum of hotness and a
maximum of coldness.   We speak of things being boiling hot
or ice cold, as though nothing could be hotter or colder.   That

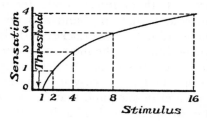

FIG. I—WEBER'S LAW FOR SENSATION

these expressions were not always mere figures of speech was
made evident to the writer by a little experience he had while a
student in Germany.   He was discussing high temperatures
with a friend, and the landlady of his boarding house happened
to be listening in.   As the temperatures rose higher and higher,
she grew more and more restive until at last, when it was mentioned that the melting point of platinum was 1700 degrees
centigrade, she burst out with "Ugh!   There aren't so many
degrees."   "How many do you think there are?" we asked.
"At most, two hundred," she replied.   Doubtless even at that
she was stretching a point, for water boils on the centigrade
scale at 100 degrees.

Modern psychologists have reduced the relation between
sensation and stimulus to mathematical law.   They find that,

as the stimulus is successively doubled, the sensation increases only additively. This is known as Weber's law. If we plot the stimuli horizontally, and the corresponding sensations vertically, we obtain the curve of Figure 1. The power of discrimination at any point is represented by the steepness of the curve, for when the curve is steep, it means that a small increase in stimulus will produce a large increase in sensation, and will therefore be the more perceptible. Now the curve is steepest at the beginning, and becomes less and less steep as we ascend. Hence the power of discrimination is greatest for the smallest stimulus, and diminishes as the stimulus increases. Note also that the curve does not start quite at the left-hand corner of the diagram, but a little to the right of this point. That means that the stimulus must reach a certain value before any sensation is felt at all. This value is called the *threshold*. Our sensations are therefore by no means proportional to the stimuli that produce them. They have limited and variable discrimination, and a limited range.

Our temperature sense being thus unreliable, and at times even contradictory, we naturally turn to something else, to some instrument that shall be exempt from human weaknesses. But every instrument that we can devise possesses these same defects, though in less degree. For example, if a mercury thermometer is first highly heated and then greatly cooled before being plunged into a bath of intermediate temperature, it will not on the two occasions register the same temperature. Beyond the boiling and freezing points of mercury, the instrument will not register any further temperatures at all. It has a limited range. A certain difference in temperature must exist before the instrument will respond. It has a threshold. It has limited discrimination or sensitivity. Finally, who can say if its readings are proportional to the temperature, and if not, what law they follow? What *are* true temperatures, and how are we to know them if our senses fail and every instrument is doubtful? That is a big problem.

Once while the author was staying at a primitive village, he sent for a nurse of the village to attend one of his children who was ill. The nurse put her hand on the child's forehead to see if there was fever. The author proffered a clinical thermometer. It was spurned. The nurse said she could tell how hot a patient was "without no thermomiker." Who, indeed, could tell if the "thermomiker" were right? We rely upon the maker. But how does he know? We take it to a laboratory to have it tested, and there it is compared with a big thermometer, which we are told is "standard." But how does the physicist know that his "standard" is right? Perhaps he admits it is a little bit wrong. But then, he has a table of corrections. Where did that come from? Very likely he will tell us that he made it himself. Did he then compare his thermometer with some other still more "standard"? No, he never went out of his laboratory. He only examined his own instrument and made out his own corrections. At this point we hand him a dollar bill for his services, and remark: "We know this bill is good because we made it ourselves."

Now how did the physicist know that his thermometer was a little bit wrong? Certainly he didn't feel of it like the nursemaid to see how hot it was? Yet what does he know about temperature, other than what his senses reveal or some instrument indicates? How does any one know what a temperature is, better than a thermometer? How can any one tell what the thermometer ought to read when it doesn't read what it ought to?

The answer is that we make assumptions. They are not made up out of whole cloth to be sure; they are *reasonable* assumptions; they are *guided* by experience as far as that is possible; but, nevertheless, they go *beyond* experience. Science always does this, and, indeed, could not get very far if she didn't.

Bodies expand when heated. The gross fact can be ascertained with no better temperature indicators than our fingers. Suppose, for example, we dip a tube of small bore, with a bulb blown on one end and filled with mercury, into a water bath that is decidedly cold to the touch. We make a scratch on the tube at the level where the mercury stands. Now we transfer the tube to a bath that is decidedly hot to the touch, and note that the mercury stands at a higher level. It has expanded. Let us mark the new level. Now suppose that we measure the distance between the two marks with any convenient scale, say, a millimeter scale, and find that the distance is fifty millimeters. Never mind how large a millimeter is. It doesn't matter in the least. Now let us dip our instrument into two other baths, and suppose that in one of them the mercury stands one millimeter higher than in the other. We *see* that the mercury has expanded, but we *perceive no difference in temperature.* We very naturally assume that there *is* a difference in temperature even though we cannot feel it, and say that it is below the power of discrimination of our senses. This is a very *reasonable* assumption, but it is an assumption none the less. If it is a fact, then we have already in this crude instrument a means of sharpening the senses. It can indicate temperature differences which are less than we can perceive. Or, rather, we have translated our rather dull tactile sensations into the much sharper visual ones.

Shall we make a further assumption? Shall we assume that the rise in temperature that produced an expansion of one millimeter is just one-fiftieth of the rise in temperature that produced an expansion of fifty millimeters? In short, shall we assume that the expansion of mercury is in exact proportion to the rise in temperature? That would be extremely convenient if true, for it would enable us accurately to measure and compare different temperature changes. But how can we test the assumption if our senses are too dull, and we have as yet no instruments? We cannot test it directly; but indirectly we can get a rough idea of how much it may be in error, and if the probable error is not too great, the assumption may be adopted as a good enough approximation, at least until a better can be found. This, in fact, is what we do.

Let us examine the expansions of other liquids, water itself, alcohol, etc., in exactly the same way we did that of mercury. We make a little tube for each, put them all in the same cold bath, then in the same hot bath, and mark the positions of the liquids in each case. We now put them all in a bath for which the mercury height stands at twenty-five millimeters. Will all the others stand at exactly their halfway marks? Very nearly, but not quite. In the same way we may test the quarter, the three-quarter, and other marks, and shall again find near agreements. We may therefore conclude that, while perhaps none of these liquids expands in exact proportion to the temperature, none departs far from it. The assumption of proportionality cannot be far from the truth.

And so this convenient assumption is made, and thermometers are constructed in much the way we have described. Even with the crude instrument we have discussed, one can discover, and, indeed, it was early discovered with instruments but little less crude, that ice always melts at the same temperature, and that water under a pressure of one atmosphere always boils at the same temperature. So the thermometer maker plunges his yet ungraduated instrument into a mixture of melting ice and water, and marks on the stem the point at which the mercury stands. He then suspends it in the steam just over boiling water and marks the point to which the mercury then rises. The difference in temperature between the freezing and the boiling points of water thus provides a natural temperature interval, which is always the same, and always reproducible. This standard interval, as marked on his thermometer, can now be subdivided in any convenient way. If he is making a centigrade thermometer, he will mark the lower point, o degree, and the upper one 100 degrees. Then he will place the thermometer in a dividing engine and accurately divide the distance between these two marks into one hundred *equal* parts. If the thermometer is to register above or below these marks, he will add further divisions of the *same size* above or below them. If he is making a Fahrenheit thermometer, he will label the freezing point 32 degrees, and the boiling point 212 degrees, and divide the distance between into one hundred and eighty equal parts, continuing the same graduations above and below if required.

It should be noted that by this mode of construction a further assumption is made, namely, that the expansion of mercury *in a glass tube* is strictly proportional to the rise in temperature. The glass expands as well as the mercury, only less so, and what the instrument really shows is the difference between these two expansions. If different kinds of glass do not expand equally, then thermometers made of different kinds of glass will show variations. In fact they do; but the variations are very slight. From the manner of their construction, it is evident that all these thermometers will agree at the ice and steam points. The disagreements will occur elsewhere on the scale.

If a substance other than mercury is used in a thermometer, the procedure is exactly the same, as though the expansion of this substance in a glass tube were also uniform. Each substance is said to give a particular *thermometric scale*, and these differ somewhat among themselves. From among them we select the *mercury in glass scale*, and *decree* that it shall be right. We do this, not because we know that it is right and the others wrong—they are probably all wrong—but because for many reasons mercury is the most convenient substance to use. Hence the mercury thermometer is right, not because it shows true temperatures, but because we have virtually defined temperature to be what such a thermometer shows. It is like the Professor's wonderful watch, described by Lewis Carroll in his story of *Sylvie and Bruno*. The watch was always right, not because it went according to the time, but because time went according to the watch. If you turned the hands forward, you jumped into the future. If you turned them back, you slipped into the past. And if you pulled out the wonderful reversing peg, everything went backwards. It certainly would be surprising to see everything going backwards, as Lewis Carroll and so many others have described it; to see, for example, the droplets of water sprinkled all over the lawn gather themselves together and shoot back into the hose nozzle. But we would not see it. The rays of light that had entered the eye and rendered objects visible, would now leave the eye and return to the objects. All the light in the world would gather itself up and return to the sun. The stars would likewise pull in their rays and leave the universe in utter darkness.

And how does the physicist correct his thermometer? He simply makes an investigation to determine whether it has been properly constructed and marked. He puts it again in the melting ice and water, and again in the steam over boiling water, and sees if these points have been correctly marked. If not, he notes the errors and calculates the resulting corrections for each point of the scale. Then he tests whether the bore of the tube is uniform. If not, he again makes corrections for each point of the scale for the variations he finds. Then he sums up all his corrections and makes a table. It is very simple. He does not have to go out of his laboratory. He can even make his own thermometer, and if done properly according to the rules, it will be right.

When a liquid is used as a thermometric substance, the instrument is necessarily limited to the range of temperatures between the freezing and boiling points of the liquid. Indeed, only a part of this range can be used, for as either limit is approached, the expansion becomes markedly irregular, as shown by comparison with thermometers using other substances. Of all liquids, however, mercury is the best in this regard, for it freezes at −38 degrees and boils at 674 degrees Fahrenheit, having thus a maximum range of 712 degrees, a good part of which can be used.

A gas can also be used as a thermometric substance, and has the advantage of a much greater range of temperature. Air, for example, liquefies at −300 degrees Fahrenheit, while the upper limit of such a thermometer is set only by the melting or softening point of the material of which the bulb is made. The scale is determined in the same way as with liquids. The volume or the pressure of the gas is measured at the freezing and boiling points of water, and the interval between is divided into one hundred or one hundred and eighty equal divisions. The temperature is then read off on this scale in the usual way. The various gas scales were found to agree among themselves better than the liquid scales did. Hence, they were deemed better scales, and the hydrogen or nitrogen scales were made standard in place of the mercury.

Thus we came to have several thermometric scales, all of them based on doubtful assumptions, probably none of them exactly true, but some apparently better than others. No one knew as yet what temperature was in a physical objective sense, nor how it could be determined other than by its *effect* on some material substance. And whether that effect were strictly proportional or not, there was no means of ascertaining.

The establishment of the mechanical theory of heat, in 1845, at last made possible a physical conception of temperature. According to this theory the temperature of a body is proportional to the average kinetic energy of its moving molecules. But no one knew at first how to translate this idea into thermometer degrees. The task was finally accomplished by William Thomson (Lord Kelvin) in 1852. He then established what he called the *thermodynamic* scale of temperature. The degrees of this scale, being determined solely by *mechanical* units, are independent of the effects of temperature upon any material substance. In fact, by means of it we can determine how far these effects deviate from true proportionality. Thus, Thomson and his coworker, Tait, drew up tables of corrections for the other scales, by which they could all be reduced to the thermodynamic. The corrections amounted only to hundredths of a degree for the hydrogen scale and to tenths for the mercury, over the ranges for which these scales are ordinarily used. Hence they turned out to be rather better than what might have been expected from the way they were made. For nearly all purposes, even scientific ones, they are still quite good enough. And liquid and gas thermometers are still made in the way described above, corrections being applied only for the most accurate work. For everyday use, temperature still is, what for most people it has always been, simply what the mercury thermometer says.

II

THE EXPANSION
OF BODIES

THAT bodies in general expand when heated and contract when cooled is well known. Probably every one has at some time removed a glass stopper that had become wedged in the neck of a bottle, by heating the neck, or has performed some similar operation, in which advantage is taken of thermal expansion. If one tightly corks a thermos bottle immediately after the latter has been filled with a hot liquid, he is apt to find after several hours, when the liquid has cooled and contracted, that the cork has been drawn in so tightly that it is very difficult to remove. On the other hand, if he puts the cork in lightly when the bottle is filled with a cold liquid, he will find after some time, that the expansion of the contents has loosened the cork, and, if the bottle has lain on its side, the contents may have spilled. Hence, when the liquid is hot, the cork should be put in lightly; and when it is cold, the cork should be put in tightly.

The expansion of a gas when heated is far greater than that of any other body, and was the first to be observed. Hero of Alexandria in the first century A.D. heated a flask full of air, then sealed it and allowed it to cool. When the seal was broken, air was sucked in. By first inverting the cooled flask and dipping its neck under water, when the seal was then broken, water was sucked in. In this way Hero was able to measure the contraction; but he had no way of determining the temperatures. We now know that when air is heated from the freezing to the boiling point of water, its volume increases by about one third. All gases show similarly large expansions.

Hero applied the expansion of air to the production of various mechanical wonders and temple tricks. His most noted achievement of the sort was to cause the doors of a temple to open mysteriously when a fire was built upon the altar. His apparatus, concealed beneath the pavement, is shown in Figure 2. The altar was hollow, and when a fire was built upon it, the air in this space expanded and passed down into a flask below, that was partly filled with water. Some of the water was thus forced out through a bent tube into a suspended bucket, which, thus weighted, pulled open the doors of the temple in the manner shown. When the fire was extinguished, the air in the flask cooled and contracted, and the water was sucked back into it. The bucket being thus lightened, the doors were pulled shut by a counterweight attached to the ropes that disappear in the drawing under the floor of the temple.

FIG. 2—HERO'S TEMPLE TRICK

Galileo Galilei appears to have been among the first to suggest that the expansion of a body could be used to indicate changes in temperature, and that air, on account of its considerable expansion, was very suitable for this purpose. In 1592 he made the world's first thermometer. It was an air thermometer. His simple device is shown in Figure 3. It consisted of a bulb having a long narrow stem projecting downward into a dish of mercury. Before dipping the stem into the mercury, the air in the bulb was slightly heated, so that, on

cooling, mercury was drawn part way up the stem. Thereafter, variations in the surrounding temperature produced
variations in the height of the mercury column. In this upside
down thermometer, what we call *low* temperatures thus read
high, and high ones low; so that, had this form of instrument
persisted, our ideas of high and low as applied to temperatures
might have been reversed.

In this instrument, the air, of course, not the mercury, was
the thermometric substance. The mercury served merely as
an index to show changes in the volume of the air in the bulb.
On account of the open dish, changes in the volume of the
mercury produced little effect. The readings were affected,

FIG. 3—GALILEO'S AIR THERMOMETER

however, by changes of atmospheric pressure, so that readings
taken on different days were not comparable. But the existence of this pressure and its variations were not then known.
The barometer was yet to be invented by Galileo's pupil
Torricelli.

This thermometer had no graduations. The mercury height
was simply measured by applying a scale. Hence the instrument could scarcely have had any other use than to detect
changes in temperature smaller than the senses could perceive.

Otto von Guericke, the inventor of the air pump, constructed
a huge thermometer of this type. Figure 4 is reproduced from
his *Experimenta Magdeburgica*, written in 1672. The great globe
at the top was the thermometer bulb filled with air. Instead of

mercury, alcohol was used as an index substance, and was contained in a U-tube, as shown in the left-hand view of the instrument from which the scale cover has been removed. A little float rode on the surface of the liquid in the open arm of the U-tube, the other arm of which connected with the air bulb.

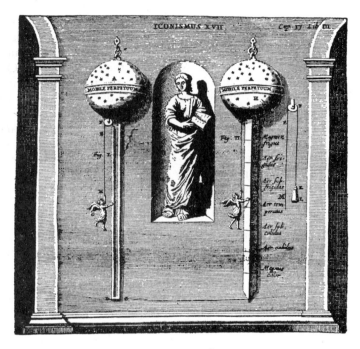

FIG. 4—OTTO VON GUERICKE'S THERMOMETER

From the float a cord passed up over a pulley, and at the other end was suspended a little angel, by whose movements up and down the changes in temperature were indicated. This thermometer had a scale of a sort, as the right-hand view shows. The divisions, rather large, were not numbered, but were provided with various names, ranging from "greatest heat" to "greatest cold." Von Guericke's directions for setting the scale are interesting. One was to select an evening in, the autumn just before the harvest, when the temperature was medium, and then draw air out of the globe, by means of a stopcock provided for that purpose, until the angel hung at the middle of the scale, which is marked "temperate heat." Some of our household thermometers to-day are still marked with temperate heat, summer heat, blood heat, etc., in addition to the numbered scale, for the benefit, perhaps, of those whose thermometry has not advanced much beyond that of von Guericke.

One may ask what the use of this huge instrument could be, which only told whether it was very hot, or moderately hot, etc., in the judgment of whoever set the scale. Von Guericke tells us enthusiastically. By means of it, he says, one can keep track of the temperature of every day in the year, and actually tell which was the hottest and which was the coldest day. One can tell whether this winter is colder or warmer than last winter, and in the course of time, I suppose, could decide the perennial question as to whether our present winters are not what they used to be twenty years ago. That would indeed be useful. But all the resources of the weather bureau seem unable to accomplish it even to-day.

FIG. 5—A
FLORENTINE
THERMOMETER

Galileo observed, also, the expansion of liquids, which is much less than that of gases. Alcohol expands one-third as much as air; water, at 70 degrees Fahrenheit, one twenty-fifth as much as air. The others fall in between. Galileo also suggested the expansion of alcohol as a temperature indicator, and around 1640 spirit thermometers much like our present instruments began to appear. The most widely used instrument of this type was the famous "Florentine Thermometer," illustrated in Figure 5. In making it, the bulb was filled with alcohol and heated until the stem was completely filled. The top was then sealed off. On cooling, the liquid descended, leaving a vacuum above it, except for the vapor of the liquid. Being thus hermetically sealed, these instruments were not affected by changes in the atmospheric pressure. Being small and light, they were transportable, and could be inserted into small places or into liquids whose temperature it was desired to measure. The divisions were marked by enamel beads fused onto the stem, and represented thousandths of the volume of the bulb. The user was thus left to interpret for himself the meaning of the readings as regards temperature. These instruments continued in use for about a century.

One disadvantage of alcohol as a thermometric substance is that it wets the tube. Consequently, when such an instrument is quickly cooled, the liquid adhering to the walls continues to drain down for some time. In 1693 Halley suggested that mercury be used, and such thermometers were soon made. Although the expansion of mercury is about one-sixth that of alcohol, the fact that it does not wet the glass makes possible the use of much smaller tubes, and this offsets the lesser expansion. The finer the bore, and the larger the bulb, the more sensitive the thermometer. A given change in volume of the mercury will then produce a greater change in the length of the mercury thread.

All the early thermometers suffered from the defect that they had no common easily reproducible scale. Different thermometers read the same temperature differently, just as different radio dials to-day read the same wave length differently. All comparisons of temperature had to be made with the same instrument. Nevertheless, with these thermometers some important discoveries were made—among others, that liquids have fixed boiling and freezing points, so that the readings of different thermometers could be compared by observing where these boiling and freezing points occurred on their scales. In 1665 Boyle suggested that all temperatures be measured up from the freezing point of water. In 1693 Halley suggested they be measured down from the boiling point. Either of these suggestions, however, still leaves the *size of the degree* undetermined. In order to establish a common unit of temperature *difference* that will be independent of the make of thermometer, it is necessary to select *two* naturally fixed points, and make the unvarying interval between them the yardstick of temperature measurement. The interval can then be subdivided in any convenient fashion.

The first to embody this principle in the construction of a thermometer was Fahrenheit in 1724. He used for his lower fixed point a mixture of water, ice, and sal ammoniac, which gave the lowest then-known artificial temperature. This he took as the zero of his scale. For the upper fixed temperature, he used the boiling point of water, which he called 212 degrees. The distance between the two marks on his thermometer stem, he divided into 212 equal parts, thus producing the prototype of the instrument so much used to-day.

It was soon found, however, that the sal ammoniac mixture did not give a reliably fixed temperature. In 1730 de Réaumur substituted the freezing point of water for the lower fixed temperature, and retaining the boiling point for the upper one marked them 0 degree and 80 degrees, respectively, with eighty equal divisions between. This scale is still much used in Continental Europe. In 1742 Celsius devised a scale in which

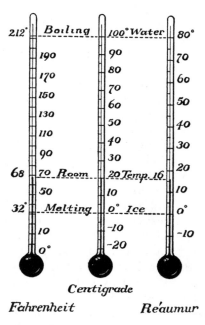

FIG. 6—COMPARISON OF THERMOMETER SCALES

he used the same fixed points, but labeled them 0 degree and 100 degrees, respectively, with one hundred equal divisions between. This is the centigrade scale, now used in scientific work throughout the world. The Fahrenheit scale is now also fixed by the freezing and boiling points of water. But the lower point is marked 32 degrees, and the interval between it and the boiling point is divided into one hundred and eighty equal parts, as explained in the preceding chapter.

The relations between these three scales are shown in Figure 6. Since between the freezing and boiling points of water there are 180 Fahrenheit, 100 centigrade, and 80 Réaumur degrees, the Fahrenheit degree is five-ninths of the centigrade, and four-ninths of the Réaumur degree. To reduce centigrade or Réaumur temperatures to Fahrenheit, then, we must divide by 5 or 4 respectively and multiply by 9. Then since the Fahrenheight zero is 32 degrees below the zeros of the other scales, we must add 32 degrees to the result. For the converse conversion, we must first subtract 32 degrees, then divide by 9, and multiply by 5 or 4.

Scientific men are inclined to look down upon the Fahrenheit scale, and to feel that it has no further excuse for existence. But for common use I think it has some advantages. People are always exaggerating temperatures. If the day is hot, they add on a few degrees; if it is cold they deduct a few. No one ever gives the air temperature to a fraction of a degree, but only to whole degrees. Now on the Fahrenheit scale, on account of the small size of its degrees, these whoppers and inaccuracies are only about half as big as they are on either of the other scales.

Solids expand still less than liquids. The volumetric expansion of aluminum, which is one of the highest, is about half that of water at 70 degrees; that of iridium, the metal used on fountain pen tips and one of the least expansible, is about one-fourth that of aluminum. The expansion of platinum is about one-third that of aluminum.

Since we deal with solids mostly in the form of rods, beams, or wire, the changes in length are more important than those of volume. The fraction of its original length by which a bar elongates, when heated one degree, is called the *coefficient of linear expansion*. It is hence the *rate* of expansion—the expansion per unit of length and per degree rise in temperature. To get the total elongation of a rod, one must multiply the coefficient by the length of the rod and by the rise in temperature. Thus the coefficient for aluminum in metric units is 0.000025. To find the elongation of a rod 100 centimeters long when heated 100 centigrade degrees, we multiply twice by 100 and obtain 0.25 centimeters. Thus, a rod of aluminum, when heated from the freezing to the boiling point of water, will expand about one-tenth of an inch per yard. For a body that expands equally in all directions, the volumetric coefficient is three times the linear.

In all accurate work, measuring scales must be corrected for linear expansion. Wood expands much less than metal, but swells considerably with moisture, and in a very uncertain fashion. Hence accurate measuring scales are always made of metal. Though the expansion is greater, the rate is known, the temperature is easily measured, and corrections can be applied.

It is a curious thing about alloys that their physical properties do not always fall midway between those of their ingredients, but sometimes much above or below any of them. The co-efficients of expansion of steel and of nickel, for example, are

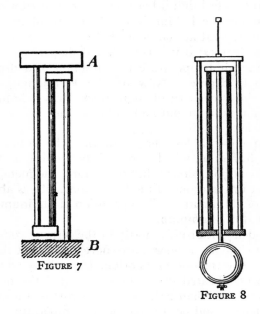

FIGURE 7

FIGURE 8

FIG. 7—TEMPERATURE COMPENSATION
FIG. 8—HARRISON'S GRIDIRON PENDULUM

about the same and equal to 0.000013. If these two metals are mixed in varying proportions, the expansion of the alloy is always less than that of either ingredient, and with 36 per cent nickel reaches a minimum that is one-fourteenth of the expansion of either ingredient alone. Alloys of nickel, steel, and other ingredients can now be made, whose expansion is practically nothing, or even negative. That is, a metal can be produced which actually contracts, though very little, when heated. These alloys are called "invar." They have been used for measuring scales, for pendulum rods, and for other purposes.

Although linear expansions are small, when pieces of great length are used, the *total* elongations may become considerable. For this reason, rails must not be laid closely end to end, but with little gaps between them to allow for expansion. For a rise in temperature of 100 degrees Fahrenheit, a thirty-foot steel rail expands about one quarter of an inch. In winter we may hear the wind singing through the telegraph wires in rather high pitched tones, for the wires are then taut. In summer we do not hear these high notes, for the wires are then slack.

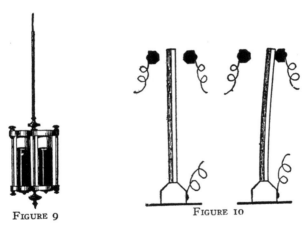

FIGURE 9   FIGURE 10

FIG. 9—MERCURY COMPENSATED PENDULUM
FIG. 10—BENDING THERMOSTAT

They must be strung, in fact, with sufficient sag to allow for the contraction in winter. The Eiffel Tower is six inches taller in summer than it is in winter when the thermometer may be 90 degrees lower. A long steel bridge span must have at least one end set on rollers to allow for expansion and contraction. If one observes the pipes laid along the walls of the New York subway or elsewhere, he will notice at intervals a crook shaped like the Greek letter Ω. This is to allow for expansion and contraction which simply close and open the crook. Where long rods or wires are used to operate signals or switches, they are interrupted at intervals by a complicated system of levers, which compensate for temperature changes which would otherwise move the signals.

Temperature compensation can be very exactly effected by using two metals of different expansibilities. Suppose it is desired to connect the two pieces A and B, Figure 7, in such a way that the distance between them shall remain invariable. We can connect them by means of the zigzag rods shown. If the expansibility of the middle rod is about twice that of the outside ones, then the upward expansion will be equal to the downward. This principle was applied by John Harrison in 1726 in his "gridiron pendulum" for clocks. This consisted of a pair of such zigzags, in order to obtain a symmetrical suspension, as shown in Figure 8. If compensation is not applied, the clock will run more slowly in hot weather because of the longer pendulum.

Another way of applying the same principle is shown in Figure 9. Here the pendulum bob consists of two little tubes partly filled with mercury. The upward expansion of the latter compensates for the downward expansion of the pendulum rod, thus keeping the effective length of the pendulum the same at all temperatures.

Another way of utilizing two metals of different expansibilities is to solder or weld two strips of them side by side. When the temperature changes, the different expansions cause the compound rod to bend. This device is much used in thermostats for temperature control. For example, in an automatic water heater, when the water reaches a certain temperature, the thermostat will bend over, and close an electric circuit as shown in Figure 10. This shuts off the gas. When the water cools a certain amount, the thermostat bends the other way and closes another contact which turns on the gas. The device is used also for small, flashing electric signs, the heating in this case being done by the electric current itself. By a device of this sort, a considerable movement can be produced by a small change in temperature. It is therefore more sensitive than one that depends upon direct elongation and contraction.

This method was also employed by Harrison in 1735 to compensate the balance wheel of a chronometer. A modification of his method is now used on all fine watches. A modern compensated balance wheel is shown in Figure 11. The rim consists of two branches, one end only of each being fixed to the opposite ends of a single spoke. These branches are composed of two metals, the more expansible one being on the outside.

Consequently, when the temperature rises, the arms bend inward. This compensates for the increase in inertia, and consequent slowing of the watch that would result from the expansion of the wheel. The elasticity of the hairspring also decreases as the temperature rises, and this would likewise slow the watch, so that the bending inward must be sufficient to compensate for this effect as well. The little weights that are screwed into the rim will be found to be quite irregularly spaced.

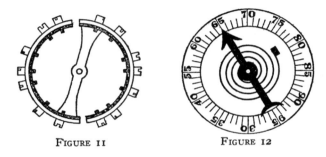

FIGURE 11   FIGURE 12

FIG. 11—COMPENSATED BALANCE WHEEL
FIG. 12—METALLIC THERMOMETER

This is because, when adjusting the compensation, the watchmaker moves them toward or away from the ends of the arms, according as the compensation needs to be increased or decreased.

The expansion of solids being small, it is not very suitable for measuring temperature. However, by coiling up a thin strip in the manner of a hairspring, a considerable length can be obtained, which will coil up or uncoil as the temperature changes. This can be made to move a needle around a dial, as shown in Figure 12. Thermometers have been made in this way. They have the advantage that the dial can be read across a room, so that it is not necessary to get up close and hunt for an almost invisible mercury thread.

A glass vessel breaks when hot water is poured into it, because the interior expands before the heat can be conveyed through the poorly conducting material to the outside. Chemists' glassware is made very thin on this account. While mechanically weak, it is thermally strong. Porcelain is about one-third as expansible as glass, and so stands the heat better. We have

no fear about pouring boiling water into a teacup. Pyrex glass
is still less expansible, and will stand the heat of an oven. The
most remarkable substance in this regard, however, is fused
quartz, whose expansion is almost negligible, being twenty
times less than that of ordinary glass. A red-hot vessel of this
substance can be plunged directly into cold water without
cracking. Unfortunately, however, the substance is very
costly.

Like many familiar sayings, the assertion that all bodies
expand when heated, is not true. Some contract. We have
already mentioned invar, certain grades of which contract.
Many bodies do not expand equally in all directions. Notably,
wood expands much more along the grain than it does across.
On the other hand, when it soaks up moisture, it expands mostly
across the grain. Many crystals expand unequally in different
directions, and some even con-
tract along certain axes.
Stretched rubber contracts
when heated, as may be shown
by suspending a small weight
by a long rubber band and
bringing a hot flatiron near the
band. The weight will then
rise. Iron expands continuously
up to 1650 degrees Fahrenheit,
a bright red heat, then contracts
a bit, after which the expansion
is resumed as the temperature is further raised. The reason is
that at the temperature mentioned, the *recalescence* point as it
is called, a change takes place in the arrangement of the molecules in the iron crystal, and the new arrangement is more
compact than the old one. Many substances have several of
these recalescence points. Alloys show anomalies in the
neighborhood of the melting point of each ingredient. Finally,
no substance expands at a uniform rate, but usually the rate
increases as the temperature rises.

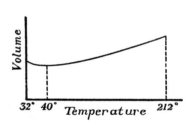

FIG. 13—THE EXPANSION OF WATER

Of all substances, liquids show in their expansions the greatest departures from uniformity. Water contracts as the
temperature is raised from the freezing point to 40 degrees
Fahrenheit, at which point it reaches a minimum volume or a
maximum density. From there on, it expands up to the boiling

point. The manner of its expansion is shown in Figure 13, in which the volumes are plotted vertically and the temperatures horizontally. The expansion is represented by a *curve*, which is concave upward. This shows that the positive rate of expansion increases continually from the freezing to the boiling point. If the expansion were uniform it would be represented by a straight line.

All liquids, when measured on the gas or thermodynamic scales of temperature, show similar upward curving expansion lines. Not all show a minimum volume like water, but in many cases, if the expansion line is continued downward—*extrapolated*—below the solidification point, it shows that such a minimum volume would have been reached had the substance not meanwhile solidified. Indeed, it is sometimes possible to prevent solidification, to undercool the liquid, as will be explained later, and reach this point.

Mercury is no exception to this rule. When compared with the thermodynamic scale, it too shows an upward curving expansion line, though fortunately its curvature is very slight. When we use mercury to measure temperature in the manner already described, we arbitrarily make its expansion line straight. All the others become curved in comparison. If we make alcohol standard, then its expansion is made straight, and, compared with it, that of mercury becomes curved the other way, and all the others curved one way or the other. Hence no liquid is, as we have pointed out, a perfect temperature measurer.

But solids also show nonuniform expansions, and oftentimes many more or less sudden and considerable irregularities such as recalescence points. Even gases are not quite uniform, so that, in the end, no material substance is a perfect temperature measurer. Only Kelvin's thermodynamic scale meets this requirement.

III

THE PERFECT GAS

THE expansion of a solid or of a liquid when heated is a comparatively simple matter, because it depends only upon the temperature. But the volume of a gas depends also upon the pressure. Liquids and solids, to be sure, are also affected by pressure, but enormous changes in pressure are required to produce very small changes in volume. Hence these effects can usually be neglected. But gases are very compressible and the pressure must be taken into account.

As the mathematician would say, the volume of a gas is a function of *two* variables, instead of only one, as is the case with a solid or a liquid. We can study the separate effects of either variable by keeping the other one for the time being constant. We shall first study the relation between volume and pressure when the temperature is kept constant.

This was first done in 1660 by Robert Boyle in a paper called "New Experiments Physico-Mechanical touching the spring of the air and its effects," and by Mariotte in 1676. They found, very simply, that when the pressure is doubled the volume is halved, and when the pressure is halved the volume is doubled, and so on. In short, the volume varies inversely as the pressure, so that the product of the two is constant. This is known as Boyle's or Mariotte's law.

If we plot this relation graphically, by laying off the pressures on a vertical scale, and the corresponding volumes horizontally, thus making a *pv* plot as it is called, we obtain the curve of Figure 14. This curve is known mathematically as an equilateral hyperbola. It has this property—that if from any point *P* on the curve perpendiculars are dropped on the two axes, the area of the rectangle formed by them and the portions of the axes they intercept is constant. For the two perpendiculars represent the pressure and the volume corresponding to P, as shown in the diagram, and the area of the rectangle is equal to their product, which by Boyle's law is constant.

We can draw some curious conclusions from this curve. Suppose the point P to slide down the curve indefinitely toward the right. The rectangle then becomes continually longer and thinner, representing the increase in volume and corresponding decrease in pressure of the gas. But the area of the rectangle remains always the same, for it is equal to the constant product

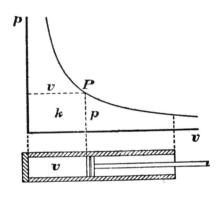

FIG. 14—COMPRESSION OF A PERFECT GAS

of p and v. However long the rectangle becomes, it must still have some width. This is equivalent to the assertion that a gas expands indefinitely, that if introduced into a vacuum it will eventually fill the whole space, even if this be as large as the universe; and it will still exert a pressure on whatever boundaries there may be; that is, no matter how much the gas may have already expanded, it will still possess the power of further expansion. Thus the curve approaches the volume axis *asymptotically*: it gets nearer and nearer but never touches the axis—unless we grant that it does so at infinity. The rectangle then becomes of infinite length and infinitesimal width, but still has the same unalterable area. We leave it to the mathematicians to figure that out—if they can.

On the other hand, suppose we push the point P up the curve by compressing the gas. The laws tells us that no matter how many times we double the pressure, we shall always halve the volume, obtaining successively one-half, one-quarter, one-eighth, . . . of the original volume—a geometrical series that ends (after an infinite number of terms) in nothing. According to the law, then, the volume of a gas can be indefinitely reduced; it can be squeezed out of existence.

We cannot believe that any real gas will act in either of these ways. However well Boyle's law may represent its behavior for middling values of pressure and volume, it cannot hold for extreme values. Somewhere the law must break down and be replaced by other laws. Boyle's law, in fact, assumes that a gas is all spring and no substance, that it is nothing but a pressure and a volume connected by a mathematical relation. But a real live gas is composed of molecules, of bits of what we may call solid matter. The pressure on the walls of a container is the summation of the individual pushes of all these separate particles. So long as the gas is fairly dense, the pressure will be sensibly uniform, and we cannot detect its spottiness. We can then treat the gas *en masse*, as though it were a continuous substance, or, if you prefer, as if it were composed of an infinite number of infinitesimal molecules. Then, and then only, can we lay down a rule like Boyle's for its *mass* behavior. But when the gas is greatly attenuated, so that its molecules are far apart, the pressure becomes a flickering and indefinite thing, to which no mass rule can be applied. A different and more complicated pattern of behavior must then replace the simple rule of Boyle.

Compression of a gas consists in pushing its molecules closer together. If the molecules have finite size, and we believe they have, this sort of compression can go on only until the molecules are brought into contact. We would not nowadays insist that compression must then absolutely stop. We no longer believe the molecules to be absolutely hard little pellets of unalterable size, as was once supposed. But it is likely that compression of the molecules themselves will be much more stoutly resisted than the mere pushing of them together. At this point, then, a different law of compression must supersede that of Boyle; and the change may be quite abrupt, like banging into a stone wall. In short, the substance should then act more like a liquid or a solid than a gas, in that an enormous increase in pressure will produce only a trifling decrease in volume. We cannot believe, then, that an actual gas can be squeezed out of existence. There must be a minimum volume that can only be further reduced by shattering the molecule.

Nevertheless, Boyle's law is very closely followed by many gases over wide ranges of pressure and volume. From this we may at once conclude that their molecules are very minute, exceedingly numerous, and ordinarily quite far apart—for they do not make their presence evident until very high pressures or enormous volumes are reached.

Indeed, for more than a century and a half Boyle's law was held to be absolutely true, and any doubts about it were regarded as a sort of scientific heresy. Nature was supposed to be essentially simple, and her laws to be the simplest possible. No simpler law than that of Boyle connecting the pressure and volume of a gas could be imagined. Hence it was concluded that it *must* be true. And when the first evidences of deviations from it at high pressures came from Despretz in 1827, and from Dulong and Arago in 1829, they were laid to experimental errors. But in 1847 Regnault, by very accurate experiments in which he compressed the more common gases up to 125 atmospheres (1875 pounds per square inch), established beyond doubt the existence of definite and progressively increasing deviations as the pressure rose.

And so the dogma of nature's simplicity received a severe body blow; and it has been badly pommeled ever since, until now a complete knock-out seems to have been achieved by Einstein.

In fact, nature is furiously complicated. But man's capacity is limited. Hence he seeks the simplest formulae. Although no part of the behavior of any gas conforms exactly to Boyle's law, it is convenient to use this law as a first approximation, and to describe the real ways of gases by their departures from it. Luckily this first approximation is frequently so close that no further corrections are necessary. The method is a common one in science. Thus, the astronomer assumes as a first approximation that the path of a planet is a perfect ellipse, although he knows that perturbations exist.

And so we have created the fiction of a *perfect gas*, which is simply one that obeys Boyle's law exactly for all pressures and volumes. Such a gas sets a standard of behavior for all other gases. It is always a gas. It cannot be liquefied. Its properties cannot in the least be altered by any amount of expansion, or by any excess of pressure. But obedience to Boyle's law is not the only demand that we shall make of it.

Having now settled the relation between pressure and volume, at least for a perfect gas, when the temperature is kept constant, we proceed to investigate the effects of a temperature change when either the pressure or the volume is kept constant. Often it is not possible to keep the pressure constant throughout the whole of an experiment, but it suffices to bring it back at the end of the experiment to the value it had at the beginning. This requirement was met in Hero's early experiments. The air was heated in an open flask, and consequently was at atmospheric pressure when the flask was sealed. When the cooled flask was inverted and opened under water, the liquid flowed in until the pressure of the air within was again one atmosphere. This amounts to measuring the contraction under constant pressure. But during the cooling, the pressure of the air in the sealed flask must have diminished. Otherwise the water would not have been sucked up. Hence when the *volume* is kept constant, a change in temperature produces a change in pressure.

A gas therefore has two temperature coefficients. One expresses the change in volume when the pressure is kept constant—this is the ordinary coefficient of expansion, such as we used for liquids and solids; the other expresses the change in pressure when the volume is kept constant. They are called the *volume* and the *pressure coefficients*, respectively.

By a method substantially the same as Hero's but more refined, Gay-Lussac in 1802 measured the two coefficients for the more common gases. He obtained the surprising result that all these gases expanded equally. Also their pressures increased equally when they were heated at constant volume. And finally the volume and pressure coefficients were equal. Thus each gas, when heated from 0 to 100 degrees centigrade at constant pressure, expanded by 37 per cent of the volume it had at 0 degree. Similarly, if heated at constant volume between the same two temperatures, its pressure increased by 37 per cent of its zero degree value. He found also that both the volume and the pressure increase between these limits was uniform, as nearly as he could determine with a mercury thermometer.

These relations are now called Gay-Lussac's law, though in England they are frequently credited to Charles, because the latter claimed to have known them as early as 1787, although he published nothing about them. If they are true, then, if we plot the volume against the pressure—a *vt* plot as it is called—

the expansion of a gas at constant pressure is represented by a straight up-sloping line as in Figure 15. The part *FB* represents the expansion from o to 100 degrees as measured by Gay-Lussac. At *B* the volume is 37 per cent greater than it is at *F*, or, using the more accurate figure of Regnault, 36.6 per cent greater. The expansion per degree is then 0.00366. Expressed as a common fraction, this figure is 1/273. A gas expands, then, 1/273rd of its o-degree volume for every degree

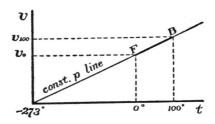

FIG. 15—EXPANSION OF A PERFECT GAS

centigrade it is heated. Conversely, it contracts by the same fraction for every degree it is cooled. At this rate, then, if the gas is cooled to − 273 degrees, it will contract 273/273rds of its o-degree volume, in other words, to nothing. The downward prolongation of *FB* must therefore hit the zero of the volume axis at − 273 degrees, and since a gas cannot contract to less than nothing, this would indicate that we here reach the greatest possible cold—the *absolute zero*.

Since the pressure change, according to Gay-Lussac, is the same as the volume change, the very same diagram will represent this case also by merely interchanging *p* and *v*. It then becomes a *pt* plot, and shows that when a gas is cooled at constant volume to − 273 degrees, the pressure is also reduced to nothing.

We cannot, of course, believe that the volume of a real gas can be reduced to nothing by any amount of cooling, any more than it could be done by any excess of pressure. Gay-Lussac's law therefore cannot hold at the extremes any more than Boyle's law can, and for the same reason. A real gas is composed of molecules of finite size, and has a minimum volume. There is no difficulty, however, in believing that the pressure of a gas can be reduced to zero by sufficient cooling. If, as we believe, the pressure is due to molecular motions, the energy of which depends upon the temperature, then when the temperature is reduced to *nil*, the motions will cease. In fact, the pressure might then be reduced to less than zero, or become negative, that is to say, it would become reversed into a contractive force. For if, as we also believe, there are forces of attraction between the molecules, then when their motions cease, they would gravitate together without the help of an external pressure.

The validity of Gay-Lussac's law was also tested by Regnault with great care and exactness. He found that the permanent gases (so-called because once believed to be nonliquefiable), air, oxygen, nitrogen, hydrogen, followed the law very closely over considerable ranges of temperature. Of course the mercury thermometer could not be used in these experiments, not only because of its limited range, but also because of the nonuniform expansion of mercury. Instead Regnault used one of the gases itself as a thermometric substance, and thus compared the expansions of the gases among themselves. Incidentally he greatly advanced the art of gas thermometry. He found the volume and pressure coefficients of these gases to be all very nearly the same, very nearly constant, and very close to the value found by Gay-Lussac. It is a remarkable fact that if the coefficients for these gases (to which may now be added helium and argon) are measured between 0 and 100 degrees centigrade and then either the constant pressure or the constant volume lines thus determined are prolonged downward, after the manner of Figure 15, they all strike zero volume or zero pressure within a fraction of a degree of − 273 degrees, and this despite the fact now known that all these gases liquefy and solidify before reaching this temperature. At or near this temperature, then, must lie the absolute zero. This conclusion has been confirmed by Kelvin's thermodynamic scale, which has placed the absolute zero at − 273.13 degrees centigrade or − 459.6 degrees Fahrenheit. Temperatures measured up from this point are called *absolute temperatures*, and may be given in centigrade or in Fahrenheit degrees. Thus the absolute temperature of the freezing point of water is 273 degrees centigrade or 492 degrees Fahrenheit.

Gay-Lussac's law then is followed closely but not exactly by

the so-called permanent gases. It is followed less closely by the
more easily liquefiable gases. We therefore treat it in the same
way that we did Boyle's law, as a convenient first approximation, and add it to the requirements of our perfect gas. According to this law too, a perfect gas is always gas. It never
liquefies, however much it is cooled. As we may see from
Figure 15, the volume increases in exact proportion to the
absolute temperature when the pressure is kept constant, and
the pressure increases in exact proportion to the absolute temperature when the volume is kept constant. A perfect gas
would therefore be a perfect temperature measurer, and either
the volume or the pressure change could be used for this purpose. Since the permanent gases follow the law closely, they
are *good* temperature measurers, much better than any liquid
or solid.

A perfect gas, then, must obey the laws both of Boyle and of
Gay-Lussac. The two laws can be combined in the one statement, that the product pv is proportional to the absolute temperature. If we give the temperature a particular constant
value, then the product pv is constant as in Boyle's law, and we
obtain a curve like that of Figure 14. If we give the temperature a different, say a higher constant value, we get another
similar curve, but situated farther out from the axes. Four
such curves are shown in Figure 16. Since the temperature is
constant along each of these, they are called *isothermals*, or lines
of equal temperature. Boyle's law is therefore the law of the
isothermals of a perfect gas.

In this figure are shown also a constant pressure and a constant volume line. The former is horizontal because, since the
pressures are measured upward, this line must maintain a
constant height above the base line, which represents zero
pressure. The latter is vertical because volumes are measured
by horizontal distances from the left-hand axis, which represents zero volume. It may be noted that the three isothermals,
which are drawn for 273 degrees, 546 degrees, and 819 degrees
absolute, which numbers are in the proportion one, two, and
three, cut off equal intervals on these, and on any other, constant pressure or constant volume lines. This is required by
Gay-Lussac's law.

It is evident that the state or condition of a gas depends upon
three variables, p, v, and T. We have used three plots, the

pv, *vt*, and *pt* plots, to show the effects of varying two of these
quantities, while the third was kept constant. But all three
may vary together. To show all the effects of such a threefold
variation, to unite all these separate plots in a single diagram,
we must use three dimensions. We must lay off *p*, *v*, and *T*
along the three edges of a cube that meet in a corner. This
may seem a little difficult, but let us not be discouraged. The
end will justify the pains.

We already have everything we need in Figure 16. Let us
imagine a *T* axis that starts at the lower left-hand corner of this

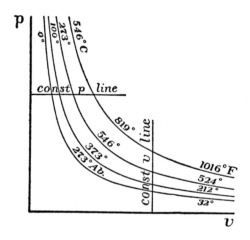

FIG. 16—ISOTHERMALS OF A PERFECT GAS

figure, and runs directly away from the reader, that is, per-
pendicular to the plane of the paper. Then let us move back
each isothermal along this axis a distance proportional to
its absolute temperature. Or, if this is too much of a strain
on the imagination, let us construct an actual model out of
cardboard.

Cut out each isothermal including its pressure and volume
axis from a separate piece, as is shown in Figure 17 for the
middle isothermal of Figure 16. Then glue it to two rectangu-
lar pieces of cardboard, which are set up at right angles to
each other like a half open book, and which represent the *pt*
and *vt* planes respectively. The isothermal is to be set back
from the front edges of these a distance proportional to its

absolute temperature, 546 degrees in this case.   Thus we might
make this distance 5.46 inches.   The front edges hence represent the isothermal for the absolute zero.   They are also the
p and v axes, while the edges along which the two boards are
joined give the receding T axis.   The pressure, volume, and
temperature of any point such as P on the isothermal are given
by the distances marked p, v, and T, on the diagram.   When
we have similarly affixed the other isothermals, we obtain a
skeleton of the model shown in Figure 18.   Since isothermals
could have been constructed for every other temperature, they
would, when all packed together in this way, form a continuous

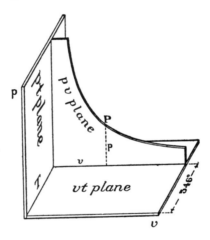

FIG. 17—CONSTRUCTION OF A MODEL

concave surface, which thins down to two sharp edges along the
pressure and volume axes, respectively.   We have cut the
model off at the freezing point isothermal, the better to show its
shape, and again at the 819 degrees isothermal; but the surface
continues of course down toward the reader to the absolute zero,
and indefinitely outward along the three axes.

Every point on this surface represents a possible combination
of p, v, and T, for a perfect gas; and every point not on the
surface represents an impossible combination of the variables
for such a gas.   We thus have a complete view of the proper
behavior of a perfect gas.

If we look down upon this model from above, we see the

constant pressure line of Figure 15, with its bit from *F* to *B* that was measured by Gay-Lussac, projected upon the bottom or *vt* plane. Every horizontal slice of the model cuts out a similar constant volume line, which runs straight to the *p* axis, showing zero volume at zero temperature.

If we look at the model from the right, we see the constant

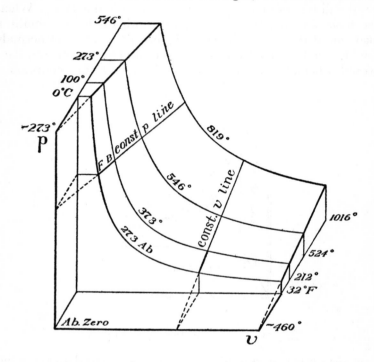

FIG. 18—MODEL FOR A PERFECT GAS

volume line, obtained from Figure 15 by interchanging *p* and *v*, projected on the *pt* plane at the left. Every vertical slice of the model running perpendicularly away from the reader cuts out a similar constant volume line, which runs straight to the *v* axis, showing zero pressure at zero temperature.

Every vertical slice parallel to the plane of the paper cuts out, of course, a *pv* plot. And so the three plots are *sections* of this single three-dimensional model.

This whole surface can be given by a single equation, $pv = RT$, in which R is a constant. It is called the *perfect gas equation*, and R is the gas constant.   Every set of values of p, v, and T that satisfy this equation gives a point on the surface, and every set that does not satisfy it gives a point that is not on the surface.   In fact, the equation merely states that the product pv is proportional to the absolute temperature, which we have already said completely defines a perfect gas.

While R is constant for any one gas, it is different for different gases.   In fact it depends upon the density of the gas, which in turn, according to Avogradro's law, depends upon the weight of the gas molecules.   Hence, if all gases were perfect, they would differ only in respect to molecular weight.   If R be given the values that correspond to the different gases, we obtain a sheaf of curved surfaces, each like the one of Figure 18, all springing from the p and v axes, like the sheets of a paper pad that are glued down along two edges, and are ruffled up at the free corner, the lighter gases being represented by the higher sheets.   Since every element will become gas at some high temperature, there are at least ninety-two of these sheets.   But there are also many compound gases.   Every gas approximates to perfect behavior if the temperature is sufficiently elevated. Hence, every gas will have a sheet, and some part of its behavior will be more or less closely represented by some part of its sheet.

IV

THOSE CALORIES

A MAN once complained to the servant at his boarding house that his room was cold. "The thermometer stands at only fifty degrees," he showed her. "Ah! But for this little room," the maid replied, "forty degrees would be quite enough."

Whether the girl was ever brought to understand the difference between temperature and heat, I do not know; but I am sure that no reader of this book would ever make a similar mistake. Yet we often carelessly use the two terms interchangeably. We say, "The heat was terrible to-day," when we merely mean that the temperature was uncomfortably elevated. Yet it is perhaps not sufficiently realized that heat is a *quantity*, just as much as a material substance is. Indeed, it was at one time supposed actually to be a material substance, an imponderable or weightless fluid, to which Lavoisier gave the name *caloric*. The thermometer does not measure this quantity. It measures only its intensity or degree of concentration.

If we place two bodies of different temperatures in contact, they assume after a time the same temperature. We say that heat has passed from the one to the other, and that the temperature of the one has fallen because it has lost heat, while that of the other has risen because it has gained heat. We *observe* the temperature changes, by means of thermometers or otherwise. We *infer* the passage of heat as the cause of them.

Two bodies at different temperatures may be likened to two vessels in which water stands at different levels, as in Figure 19. The water represents a quantity of heat, and the levels represent the temperatures. When the two vessels are put into communication, water flows from one to the other until the levels are equalized. If there is no leakage, the quantity of water that enters the one is the same as what left the other. If

the diameters of the vessels are the same and are uniform, the
level of the one rises as much as that of the other descends, and
the final level is halfway between.

But if heat quantity is not measured by a thermometer, how
then shall we measure it? The answer is—in the same way
that we measure water. And how do we measure water?
Why! We provide a vessel having a certain *capacity*, fill it to
the top, and then say, for example, we have a quart.

Let us put a kettle of water on
the stove and place a thermometer
in the water. If the fire is a gas
flame that burns at a steady rate,
we may assume that heat flows
into the water at a steady rate, and
we shall find that the rise in temperature is also uniform. Equal
amounts of heat are added to the
water in each minute, and equal
changes in temperature occur each

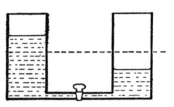

FIG. 19—COMMUNICATING VESSELS

minute. If we fill the kettle half full, we shall find that the
boiling point is reached in half the time that the full kettle
required. The half quantity of water has therefore half the
capacity for heat that the full quantity had; it took half the
quantity of heat to fill it up, so to speak, to the boiling point.
The *heat capacity* of a body is therefore proportional to the
quantity of the body, or more exactly, as could be ascertained
by more precise experiments, to its mass or weight. In a given
weight of water, or of any other substance, we have then a
definite heat capacity. It requires a definite amount of heat to
fill it from one mark to another on the thermometer scale.

In measuring a liquid, we usually start with an empty vessel
and fill it to the top. We cannot do this with heat, for we
cannot empty a body of all its heat, that is, cool it to the absolute zero. We have to start with the body already partly filled,
so to speak. We often do this in measuring liquids. Suppose
we have a graduated cylinder, as in Figure 20, already filled to
a certain mark with water, and we wish to add another quart.
We simply pour in water until the level rises to the mark that
indicates another quart. We do the same thing with heat.
Suppose we have a kilogram of water already filled with heat to
the o-degree mark on the centigrade scale, or 273 degrees

absolute. We add heat until the temperature rises to the 373-degree mark, and then say that we have added one hundred *large calories*.

The large calorie is therefore the amount of heat required to raise the temperature of one kilogram of water one degree on the centigrade scale. This is the heat unit that overweight women are so much concerned about, without perhaps knowing

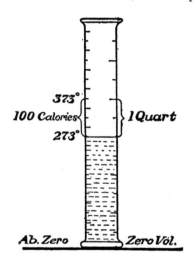

FIG. 20—CALORIES AND QUARTS

just what it means. They tell us that an egg contains 75 calories. That means that, when completely oxidized or burnt, it will develop enough heat to raise the temperature of a kilogram of water 75 degrees centigrade, or to raise about a pint and a half of water from the freezing to the boiling point. These people unwittingly eat, or perhaps abstain from eating, by the centigrade scale, but use the Fahrenheit in all their other dealings.

For scientific purposes the heat unit most used is the small calorie, which is the thousandth part of a large calorie. For engineering purposes in English-speaking countries, the British Thermal Unit, abbreviated to B.T.U., is mostly employed. This is the quantity of heat required to raise the temperature of one pound of water one degree on the Fahrenheit scale. We shall call this unit a *therm*.   It is equal to 252 small calories, or about one-fourth of a large calorie.

All substances do not have the same capacity for heat, weight for weight, that water has.   The amount of heat required to raise the temperature of one gram of any substance one degree centigrade—or one pound, one degree Fahrenheit—is called the *specific heat* of the substance.   Since the amount of heat required for this purpose in the case of water is the unit of heat, the specific heat of water is 1.   All other specific heats are therefore expressed in terms of that of water.   Being thus relative numbers, they are, like specific gravities, independent of the units employed in measuring them.   The *thermal capacity* of a body is its specific heat multiplied by its weight.   It is the total heat absorbed by the body when its temperature is raised one degree.   Thus five pounds of iron will absorb five times as much heat as one pound of iron will, when both are heated the same number of degrees.

Let us take a pound of boiling water at 100 degrees centigrade, and mix it with a pound of cold water, at, say, 10 degrees centigrade.   The mixture will assume a temperature of 55 degrees, just halfway between the extremes.   The hot water has cooled 45 degrees, while the cold water has warmed 45 degrees, and of course the heat lost by the one is the same as that gained by the other, barring any loss during the transaction. This is illustrated by our communicating vessels of Figure 19 when their diameters are equal.   The one level, as we explained, then sinks as much as the other rises.

Suppose, now, that instead of a pound of water at 100 degrees, we introduce a pound of iron at 100 degrees into our cold water.   The final temperature of the mixture will then be 20 degrees.   The iron has fallen 80 degrees, while the water has risen only 10 degrees, in receiving the same amount of heat that the iron lost.   The heat capacity of the iron is therefore only one-eighth that of the equal weight of water.   Its specific heat is 0.125.   This case would be illustrated by communicating vessels in which the diameter of one was only one-eighth that of the other.   The level in this one then sinks eight times as much as that in the other one rises.   The diameter of the vessel hence corresponds to the specific heat.

Now the diameter of a vessel might not be uniform.   That would mean that the specific heat of a substance might vary with the temperature. That has been found to be the case. The heat absorbed by a gram of water between 0 and 1 degree centigrade is not the same as that absorbed between 15 degrees and 16 degrees. Hence, in defining the calorie, we must for exact work specify the two temperatures between which the one-degree rise in temperature is to take place.

The specific heats of nearly all substances are less than that of water, and hence are expressed by decimal numbers. Solids are particularly low, all the common substances ranging from 0.031 for lead to 0.22 for aluminum (from 1/30 to 1/5). Ice, however, is an exception with a specific heat of 0.5. There are also some less common substances with high specific heats. That of lithium, for example, at ordinary temperatures is 0.84, and increases rapidly with the temperature, so that above 100 degrees centigrade, the specific heat of this metal is greater than that of water. This is very exceptional. The noble metals, gold, mercury, platinum, iridium, tantalum, tungsten, on the other hand, all have low specific heats around 0.03.

Liquids in general have higher specific heats than solids. They range mostly around half that of water. Mercury, however, follows rather the rule for the noble metals with a specific heat of only 0.033. This low thermal capacity of mercury is another quality that adapts it for thermometers. It abstracts but little heat from the body whose temperature is to be measured.

Water, on account of its high thermal capacity, is an excellent storehouse of heat. No better substance can be found with which to fill your hot water bottle. The great quantity of heat contained in the ocean, and particularly in its warmer currents, has a marked effect in equalizing the climate. On the coasts, the extremes of temperature are much less than inland. For the same reason, water is the best cooling agent. In the radiator of an automobile, or in the condenser of a steam plant, it will carry away more heat per pound than any other liquid.

When it comes to gases, we find that just as they had two co-efficients of expansion, so they have also two specific heats, according to whether the pressure of the volume is kept constant during the heating. The specific heat of a gas is very difficult to measure with accuracy, for since the heat capacity of a body depends upon its weight, and a gas is very light, an enormous volume is required in order to absorb an appreciable and accurately measurable amount of heat. The difficulty is not so great in the case of the specific heat at constant pressure, for in that case the hot gas may be allowed to stream through a coil of pipe immersed in the water of a calorimeter, the rise in temperature of the water and the fall in temperature of the gas as it passes being simultaneously measured. In this way a large volume of gas can be used. By this method Regnault in the middle of the last century made the first accurate and extensive measurements of the specific heats of gases at constant pressure.

In the case of the specific heat at constant volume, however, a fixed quantity of the gas must be confined in a closed container, and this cannot be very large. Now even the lightest possible container will be many times heavier, and will therefore absorb many times more heat than the enclosed gas will. It is necessary first to determine how much heat the container absorbs without the gas, and then how much it absorbs with the gas. The difference between these two large and nearly equal quantities gives the amount of heat the gas alone absorbs. This is like trying to weigh a feather inside a leaden casket, by weighing the casket with and without the feather. Suppose we commit an error of one part in a thousand in weighing the casket. If the casket weighs a hundred times as much as the feather, then the error in the weight of the feather will be one part in ten or 10 per cent. Because of this difficulty, direct and accurate measurements of the specific heats at constant volume were not made until 1890, when Joly invented a new kind of "differential" calorimeter for the purpose. Even to-day but few gases have been measured in this way. Fortunately, however, we have indirect ways of calculating these specific heats from other data. But this matter is too complicated to discuss here.

The specific heats of the permanent gases at constant pressure are all much the same, and range from one-half to one-eighth. That of air is 0.24. Hydrogen, however, is a great exception, having the enormous value of 3.40. This is the only substance that has a decidedly higher specific heat than water. This undoubtedly accounts for the high thermal capacity of water, which contains considerable hydrogen. The specific heat of steam is 0.47, that of ice 0.5, so that water when it becomes either vapor or solid loses about half of its thermal capacity.

The specific heat of a gas at constant volume is always less than its specific heat at constant pressure.   For example, the former for air is 0.17, that for steam is 0.34, for hydrogen 2.40. Comparing these with the corresponding values for constant pressure given above, they are all seen to be less.   The reason for this long remained a profound mystery, until J. R. Mayer in 1842 propounded the theory that heat is a form of energy, that a calorie will produce a definite number of kilogram-meters of work—a therm, so many foot pounds—and *vice versa*.   He pointed out that when a gas is heated at constant pressure, it expands and does work against the opposing atmospheric pressure.   A certain amount of heat is required to raise the temperature of a gas one degree at constant volume.   This is its real specific heat.   A certain additional amount of heat is required at constant pressure to do the work of expansion. From this additional heat, and the work done, he showed that the number of kilogram-meters produced by the consumption and disappearance of one calorie—what we now call the *mechanical equivalent of heat*—could be calculated.   This is the principle of the conservation of energy.   A year later Joule independently discovered this principle, and made the first direct experimental measurements of the mechanical equivalent.   These have been often repeated and refined by Rowland and others.   The accepted value to-day is 778 foot pounds to one British Thermal Unit.

By the work of Joule, Kelvin, Clausius, Tyndall, and others, it was further established that heat is a form of molecular agitation, or as Tyndall put it, "a mode of motion."   This is the mechanical or kinetic theory of heat.   Although it had been often proposed long before, it was only established after a long and bitter struggle with the official caloric theory.   But that is another story.

V
LATENT HEAT

ONE morning, the author of these pages complained to the cook that his coffee was not very hot. She replied: "Well, I boiled it for half an hour and couldn't get it any hotter." We feel inclined to ask, "What's wrong with this picture?" While most persons may be aware of the fact that prolonged boiling does not raise the temperature of a liquid, nevertheless it is difficult to persuade any cook that to boil vegetables "hard" does not cook them any faster or any better, than to boil them gently. Yet with a thermometer it is easy to show that no matter how fierce the fire or prolonged the boiling, the temperature of the water cannot be raised above its boiling point, 212 degrees. Now, during the boiling the fire is continually delivering heat to the water. What becomes of it all? If the kitchen is small and not very well ventilated, we shall soon find out (especially if a thermometer hangs on the wall) what becomes of it. Certainly the heat that causes the cook to grumble about the hot kitchen doesn't do the vegetables any good—nor the cook, nor those who have to listen to her, nor the one who has to pay the gas bills.

The heat supplied to the water serves merely to maintain its temperature. Except for a small amount required by the chemical changes that constitute cooking, all of it goes to make up the losses to the surroundings. If these could be prevented, and they can be greatly reduced by efficient heat insulation, such as that of the fireless cooker and the electric oven, a surprisingly small amount of heat would do the cooking. Our open pans, iron stoves, and uninsulated ovens are a disgrace to an age that calls itself scientific. The only good that violent boiling can do is to maintain a lively circulation of the food. But it is cheaper to do this with a spoon.

Let us put a kettle of ice-cold water on a steady gas fire and see how long it takes to boil. We shall find that it takes more than five times as long to boil all the water off as it did to heat the water from the freezing to the boiling point. And since the fire has burned steadily all this time, it must have delivered more than five times as much heat to the kettle during the boiling as during the heating period. Yet all this fivefold quantity of heat did not raise the temperature of the water one jot. Neither did it raise the temperature of the steam. If we place a thermometer in the steam just above the liquid, we shall find that it again stands at 212 degrees during the whole of the boiling. Only when the last drop of water has boiled off, can the temperature of the steam be raised. Then the temperature rises with sudden and alarming rapidity, and if the fire is not immediately extinguished, the kettle will burn.

That all this heat that produced no rise in temperature has not somehow gone out of existence, but is still in the steam, we can prove by getting it out of the steam again. If, instead of allowing the steam to escape and cause the cook to grumble, we had conducted it by means of a hose to a coil of pipe immersed in a large body of cold water, the steam would have condensed in the pipe and raised the temperature of the surrounding water. In fact, we could in this way recover from the steam more than enough heat to raise the temperature of five similar kettles full of water from the freezing to the boiling point, in short, all the heat that we had put in.

More exact measurements show that to boil one gram of water at 100 degrees centigrade and convert it into steam at the *same* temperature, requires 538 calories of heat, whereas to raise the temperature of a gram of water from 0 degree to 100 degrees centigrade requires only 100 calories. And conversely, by the condensation of that same steam at 100 degrees centigrade to water at the same temperature, provided there are no losses, 538 calories of heat can be recovered. Or to speak English, the conversion of a pound of water at 212 degrees Fahrenheit into steam at the same temperature requires 970 therms, whereas to heat the water from 32 degrees to 212 degrees requires only 180 therms.

This absorption by a boiling liquid of a large quantity of heat was discovered by Joseph Black in 1757. He called it the *latent heat of vaporization*. Since this heat did not raise the temperature of the liquid, but could be recovered from the steam,

and was then as competent to raise the temperature of other
bodies as heat derived from any other source, Black supposed
that it must meanwhile have existed in some sort of inactive or
latent state.

Steam at 100 degrees centigrade then, contains a great deal
more heat than water does at the same temperature. This
explains why a severe burn is produced by steam, and very little
by boiling water. The water in cooling from 100 degrees
centigrade to the temperature of the skin, 37 degrees centigrade,
gives up only 63 calories per gram. But the steam in condensing and cooling to the same temperature gives up over 600
calories.

Steam is therefore very useful for heating purposes. But
while a pound of steam contains over six times as much heat as a
pound of water at 212 degrees Fahrenheit, both being measured
up from the freezing point, the steam requires sixteen hundred
times as much space as the water. A volume of water equal to
that of the steam would contain two hundred and fifty times as
much heat as the latter. Hence, for storage of heat, water is
better. It is still the best thing to put in your hot water bottle,
because you can get so much more in a given space. Steam is
suitable for heating when there is always a direct connection
with the boiler, and new steam can be continuously supplied as
fast as the old steam is cooled and condensed. Since steam
travels very fast—fourteen hundred feet or more per second,
depending upon the pressure—heat can be delivered in this
way very rapidly.

The 538 calories per gram that water requires for its vaporization is more than is required by any other liquid. The latent
heat of mercury is only 63 calories, that of alcohol 207, of
gasoline 95 calories. It may be interesting to note that the
liquefied gases have the following latent heats: hydrogen 123,
oxygen 58, nitrogen 50, air 50, and carbon dioxide 57 calories
per gram. Although hydrogen is higher than the rest, it is not
nearly so exceptional in this matter as it is with respect to its
specific heat. And none of these figures even approaches that
for water.

A considerable amount of heat is also absorbed during a
change from the solid to the liquid state, though not as much as
during vaporization. This absorption of heat can be demonstrated in the following way. Let us fill the inside container of

an ice cream freezer with water, and having inserted a thermometer, surround it with the usual mixture of ice and salt. With one part of salt to. three of finely shaved ice, this mixture can be made to reach a temperature as low as o degree Fahrenheit, and if renewed from time to time can be kept at that temperature. The temperature of the water in the container descends gradually until at 32 degrees the descent stops. Freezing has now begun, and during the whole of this process, provided the mixture of ice and water is constantly stirred, the temperature remains stationary at 32 degrees. When at last all of the water is frozen, the temperature again descends. The ice becomes colder, and approximates finally the temperature of the surrounding freezing mixture. During the whole of the rather prolonged period of freezing, heat must have been continually withdrawn from the water by the colder surroundings. This is the *latent heat of fusion.*

If now we remove the can from the freezer and expose it to the warm surroundings, the reverse processes take place. The temperature of the ice rises until 32 degrees is reached, and melting begins. During this process, the temperature again remains stationary, and, if the ice and water are kept well stirred, will not rise until the last bit of ice has melted. Then the water warms up, and approaches the temperature of the surroundings. During the long melting period, heat must have been continually absorbed from the warmer surroundings, in fact, as exact measurements would show, the same amount of heat as was withdrawn from the water during the freezing. These facts were also discovered by Black in 1757.

The latent heat of fusion of ice is 80 calories per gram, or 144 therms per pound. Almost no substance has a higher value. Mercury requires for its melting only 3 calories per gram, lead 5, and aluminum 77.

Before the facts of latent heat were discovered by Black, it was supposed that the rise in temperature of a substance through its melting point was continuous, and that no more heat was required than for a similar rise in temperature elsewhere along the scale. If this were so, Black said, the consequences would be frightful. It would mean that when a block of ice had reached the melting point, the addition of a very small amount of heat would convert the whole mass almost at once into water. Instead of this, we observe that ice, even when exposed to quite high temperatures, melts slowly and only on the surface, and does not rise above the melting point until it is all melted. Even as things are, Black said, the melting of mountain snows is often sufficiently rapid to produce great torrents and floods. We may add that ice would be of no use in a refrigerator if a very little heat sufficed to melt it all at once.

VI

CHANGE OF STATE

DURING a change of state there is always a change in volume while the temperature remains stationary. This distinguishes it from the ordinary thermal expansion which takes place with a rising temperature. Figure 21, for instance, is a vt plot of ice, water, and steam. The proportion from 32 degrees to 212 degrees is identical with Figure 13, and shows the volume changes of water as the temperature varies between those limits. At those limits, however, the line rises vertically, showing expansions without change of temperature. At the boiling point the expansion is enormous, so great that it cannot be shown on the scale of this diagram. Hence the vertical line representing it is broken. A cubic inch of water at 212 degrees becomes nearly a cubic foot of steam, more exactly sixteen hundred cubic inches. If the height of the water line at this point were one inch, that of the steam line would be 133 feet. All liquids show similarly large expansions on vaporizing. The up slope of the steam line shows the resumption of thermal expansion with rising temperature after the liquid has vaporized.

The change in volume at the freezing point is much smaller, 10 per cent for water, and this is larger than for almost any other substance. The down slope of the ice line shows that the ice contracts as its temperature is lowered, just as any other solid does.

Although water and a few other substances expand on solidifying, most substances contract, and the volume change is usually very small. Paraffin wax, however, contracts considerably. If a pot of melted paraffin is allowed to cool, a deep depression forms at the center of the surface, and resembles the dimple at the top of an apple. This occurs because the wax contracts and, solidifying first around the sides and bottom, the surface is drawn down in the center. Bronze and brass contract very slightly. Aluminum contracts so much that good castings are difficult to make. Type metal expands slightly and, thus filling every corner of the mold, good sharp castings can be made.

Since water expands 10 per cent on freezing, a block of ice floats with nine-tenths below and one-tenth above the surface of the water. It is commonly said that an iceberg floats with one-seventh above the surface. This is partly because sea water is denser than fresh water, and partly because common ice is full of air bubbles which lighten it.

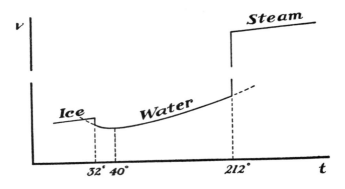

FIG. 21—EXPANSION OF ICE, WATER, AND STEAM

The expansion of water on freezing causes a good deal of damage. It bursts our water pipes, spoils fruits and vegetables by bursting their cells, cracks and disintegrates the rocks, thus promoting erosion and the removal of the soil to the sea, and a late spring frost destroys the young buds and the sap. It seems to be an unmitigated nuisance. But if water contracted on freezing, the consequences would be far worse—in fact, disastrous. In that case ice, as soon as it formed on the surface of a pond, would sink to the bottom. The warmer water on top would be continually exposed to the freezing weather, and would continually freeze and sink. Before long the whole pond would be a solid block of ice. Many of our lakes, rivers, and harbors would freeze solid. In summer, the water that resulted from the first melting would cover the ice below and hinder the warmth from reaching it. And this hindrance would increase as the melting proceeded, with the result that many of our frozen lakes and rivers would never thaw out. The polar caps would extend into and cover most of the temperate zones, the glaciers would descend deeper into the valleys, and the climate of the whole world would become so frigid that only the equatorial regions would remain habitable.

But since ice floats, it remains on the surface and hinders the heat from flowing out of the water below. And the thicker the ice the greater the insulation. The consequence is that a point is reached where further freezing cannot take place, however severe or prolonged the cold. Even in the polar regions ice never directly freezes to more than six or seven feet in thickness. Where it is thicker, it is because it has been piled up in jams, or has slid down from the land as glaciers, or because snow has fallen on it and congealed to ice. In summer the water resulting from the first melting, being heavier, seeks every crack and crevice to get under the ice. Thus a fresh surface of ice is always exposed to the melting rays of the sun. The fact that ice floats thus hinders freezing in winter and promotes melting in summer, thereby preventing great accumulations of ice, which, so to speak, store up the cold of winter, and require a good portion of the summer's heat to thaw them out.

We have seen that in general, when heat is steadily poured into a solid, its temperature rises and its volume increases continuously. But at the melting and boiling points discontinuities occur. At each point, there is a change in volume without change in temperature, and the absorption or rejection of a latent heat. These are the two essential characteristics of a change of state.

In Chapter II we mentioned that iron expands up to 1650 degrees Fahrenheit, and then contracts a bit. If a very hot piece of iron is allowed to cool, the glow gradually diminishes until this point is reached. Then it brightens up for a moment, after which the fading is resumed. It is because the iron thus seems to reheat itself a bit, that this point is called the recalescence point. There is a pause or discontinuity in the temperature drop, the release of a latent heat, and, as we saw in Chapter II, a change in volume. In short, we have here all the characteristics of a change of state, so that what happens has just as much right to this title as those larger and more conspicuous changes from solid to liquid and from liquid to gas, with which we are more familiar. We shall see that these minor changes of state are quite common. And this change at recalescence we saw was due to a rearrangement of the molecules. This is, in fact, the essential cause of *all* changes of state.

A change in molecular arrangement involves a change in molecular forces. When the molecules are separated, these forces are attractive, diminish rapidly with the distance, and at a short distance become insensible. Since we no longer conceive the atom as a solid mass of matter, but as made up of exceedingly minute protons and electrons and hence nearly all space, and the molecule is a collection of atoms, we have to adopt some convention as to what shall constitute the boundary of a molecule. When the distance between the centers of two molecules is reduced to a certain amount, the attraction is more or less suddenly replaced by a repulsion, which increases very rapidly as the distance is further diminished. We can only say that the point where this inversion of force occurs is on the boundaries of both molecules, and the two are then in contact. After all, this is all we could say even if the molecules were solid particles. Any closer approach means that the boundaries must be pushed in, the molecules themselves compressed. This is resisted in the same way that an elastic body resists compression. When the compressing force is removed, the boundaries spring back into shape, so that when two molecules collide they rebound like elastic bodies.

In the solid and liquid states, the molecules are, as we have said, in contact or nearly so. Hence in these states bodies stoutly resist compression, which would largely consist in compression of the molecules. They also resist tension, or the separation of the molecules. Hence both solids and liquids have definite volumes, which can only be altered by force, or by change of temperature. The mean distance between the molecular centers is conserved.

In the solid state, furthermore, each molecule is held to a fixed position, and any effort to displace it is resisted. If a molecule is forcibly displaced, but not too much, it springs back to its original position. Now a change in the volume of a body involves only a change in the distances between the molecules, but a change in shape requires changes in their relative positions. Since these are resisted, a solid resists change of shape. It has, as we say, *rigidity* or *form elasticity*. When deformed, if not too much, it springs back into shape.

No liquid is springy in this way. Its molecules slide easily over one another like slippery fish eggs, and none has a definite place. Hence a liquid flows. It has no definite shape, but assumes that of the containing vessel and a level surface.

Every liquid, however, offers some resistance to this sliding of its molecules. The liquid may be thin and mobile like alcohol or water and flow easily, or it may be thick and *viscous* like honey or molasses and flow sluggishly. The former will very quickly assume the shape of the containing vessel and a level surface; the latter will require more time; but however viscous the liquid, it will eventually do these things. Viscosity is hence very different from the spring of solids, which eventually *stops* the displacement. Viscosity is a species of *internal friction* which *retards* the flow, but cannot stop it. So long as the force is applied, the flow continues, though it may be very slow. Thus molasses will continue to run down hill, even if the slope be gentle, so long as there is any hill to run down.

This flowing, which is so characteristic of liquids, can be induced, under certain conditions, in solids. If a sufficient force is applied to the molecule of a solid, it may be shifted to a new position. And if the force is not so great as to tear the molecule completely away from its neighbors, that is, to break the solid, then, when the force is removed, the molecule will adopt its new position and maintain it vigorously, just as it did the old one. In short, the piece has now received a permanent deformation or *set* as we call it, and is now springy with respect to the new form as it was with respect to the old one. Thus when a steel saw has become bent, it always springs back to the bent shape.

A force sufficient to produce a permanent set is said to exceed the *elastic limit*. If such a force—insufficient, however, to break the piece—is maintained, a continual shift or sliding of the molecules over one another will be produced, and will continue as long as the force is applied. This is a flowing of the substance precisely like that of a liquid. The only difference is that it is far more highly resisted. The viscosity of a solid is thus much greater than that of a liquid.

The distinction between a solid and a liquid therefore consists not in the fact that a solid cannot flow, but in the fact that to produce a flow the force must exceed a certain amount, namely, the elastic limit. Loads up to this limit are *permanently supported*. Thus our columns and beams, if not overloaded, continue to hold up our buildings indefinitely. Solids thus possess a *limited* rigidity. Liquids, on the other hand, possess no rigidity. The smallest force will produce some flow, no matter how viscous the liquid. And the flow will continue, that is, the deformation will increase, as long as the force is applied.

In general, only a limited amount of flow can be produced in a solid without breaking it. When this amount is very small, the solid is said to be brittle; when it is large, the solid is ductile or plastic. Indeed, some solids seem to have no springiness at all, since they can be bent or twisted into any shape we desire. But they must have some, for otherwise they would not retain the shapes we give them.

Flowing always takes time. Every one knows that, in order to bend an object that is not very ductile, the job must be done slowly and carefully. The elastic limit must be exceeded, but the breaking point must not be reached. It is therefore often a ticklish business, which cannot be hurried.

Since to hold a molecule to a fixed place requires stronger forces, and hence a closer packing, than merely to hold it to a fixed distance, and some solids are less dense than their liquids, the rigidity of these solids must be due to a closer packing of the molecules in some places than in others. These strong parts run through the mass like the framework of a building. Hence a solid has structure. And the structure of a solid consists in having a place for every molecule, and every molecule in its place, except in so far as it is compelled by force to alter its place.

A liquid, instead, is structureless. No molecule has a definite habitation, but wanders from place to place. On the whole they are uniformly distributed, for which reason a liquid is *isotropic*, that is, has the same properties in all directions. Usually, also, it is clear or transparent, except when it is a metal, for all metals are opaque. A solid is usually *anisotropic* and opaque.

The structures of solids are of three sorts. In one type the molecules are arranged in a geometrical pattern, in straight lines and planes that meet at very definite angles. This is the crystalline type. Such a solid has a definite melting point, fuses or solidifies bit by bit at a constant temperature while absorbing or rejecting a latent heat. This is the type we have already described.

In the second type the structure is irregular. In places the molecules are clumped together, in others more widely separated. The fracture of such a solid shows a grain, but no crystal facets. This is the *amorphous* or formless type. It includes such substances as fats, gums, resins, waxes. These solids have no melting points. When heated, they gradually soften throughout the whole mass, becoming first pasty, and then a thick fluid, which thins as the temperature is raised.

This sort of melting is similar to what all melting was supposed to be like before the days of Black, and which he said would be so disastrous in the case of ice. If heat is supplied at a constant rate, there is no halt in the temperature rise while a large latent heat is absorbed, as is the case with crystalline solids. However, there is a slowing up in the temperature rise, an increased rate of softening over a certain temperature range, so that while a definite latent heat cannot be assigned, more heat is absorbed along this range than along an equal range elsewhere on the scale. The whole change seems to be merely one of viscosity, which becomes more rapid in this neighbourhood. The nearest approach we can make to a definite melting point is to assign the point at which this change is fastest.

In fact these substances even in their so-called solid states are more like highly viscous fluids than like true solids. For the smallest force steadily applied will produce in them a continuous yielding or flow. Thus a stick of sealing wax supported at its two ends will gradually sag in the middle of its own weight, though the amount may be perceptible only after several hours, conspicuous only after several days. Much depends upon the temperature, for the viscosity of these substances varies greatly with the temperature. Wax candles will bend over and droop on a warm day until they are almost double. But if you try to straighten one up, you will surely break it, unless you are willing to spend an hour or so at the job. A penny placed on a piece of tar will gradually sink through it and the tar will close in over it.

In this connection the author may perhaps tell a little story on himself. He owned at one time a small house in California near an abandoned oil field. The ground all about was covered with a layer of a tarry substance several inches thick. He wished to plant a garden in his back yard. But it was necessary first to remove the tar. At first he went at it energetically with a pickax, intending to break it up by main strength and stupidity. After an hour's hard work, almost nothing had been accomplished. Then he bethought himself of his physics. "This is a tarry substance," he said. "It will not yield to violence. We must use persuasion." So he drove his pickax under the edge of the tar, wedged a rock under the pick as a fulcrum, and stood on the other end. For a minute or so nothing seemed to happen. Then gradually the tarry layer rose, until at last a fine big piece broke off. In this way the whole yard was finally cleared with less fatigue than the first hour had cost. Thus it sometimes pays to know something—especially if you are lazy.

The third type of solid is still more formless than the preceding. The molecules are uniformly distributed, precisely as in a liquid, but fixed in their positions instead of wandering. Like the liquids, they are isotropic, and generally clear or transparent. A fracture is smooth, and shows neither grain nor crystal facets.

But far from being soft and plastic like the amorphous bodies already discussed, these substances when cold are among the most solid of the solids. They are extremely rigid, hard, and brittle. They permit of scarcely any flowing at all. Like the other amorphous bodies, they soften as their melting points are approached, but the softening does not begin until quite close to these points, and is then quite rapid. They can then flow tremendously, and be drawn out into long fine threads. These are the glasses, of which window glass is one variety. The nature of these substances, however, will be discussed in Chapter XII.

Finally in the gaseous states the molecules are so widely separated that they are most of the time quite beyond the effective range of their mutual attractions. They only come within that range during the brief moments of collision, but immediately rebound and separate again. They are on the whole uniformly distributed.

The mechanical theory of heat affords a ready explanation of these changes of state. According to this theory, as we have seen, heat is a species of molecular agitation, which increases with the temperature. In the solid state the molecule oscillates about its mean position. As the temperature rises, the increased knocking of the molecules against one another thrusts them farther apart. The body expands. When a certain temperature is reached, the oscillations become so large that the displacements exceed the elastic limit. The molecules can no longer be held in place. The solid structure disintegrates. The substance flows. It becomes liquid.

To tear a molecule away from its neighbors requires energy, and this is furnished by the heat supplied. The heat that is used up in this way cannot produce any increased thermal agitation, for that would mean to eat your cake and have it too. Hence occurs the considerable amount of heat absorbed without change in temperature, the latent heat, in the case of the crystalline solids, and the excess of heat absorbed during a retarded rise in temperature, in the case of the amorphous bodies. Although this heat apparently disappears, the energy it represents remains in the *potential* form. The molecules have been separated against their mutual attractions. This is like separating stones from the earth. The energy consumed in lifting them can be regained when the stones are allowed to fall upon the earth again. Hence we say that this energy has meanwhile existed in the *potential* form. We could just as well have said that it existed in a *latent* form. And so the heat energy consumed in lifting the molecules can be recovered when, by lowering the temperature, they are allowed to fall together again.

Just as it takes time to demolish a building, so it takes time to disintegrate a solid. And the time required depends upon how fast heat energy is supplied for the work. We might do our demolition job by putting all the workmen on the roof and having them break off pieces one at a time until the entire building is razed. This is how it happens with a crystalline solid. Or we might distribute the workmen all over the building, and set them to prying the stones loose until the whole building collapses. This is how an amorphous body disintegrates. Notice how a pat of butter, as it is warmed, gradually slumps.

In the liquid state, the molecular forces are no longer strong enough to hold the molecules to fixed positions, but only to a fixed average distance apart. The molecules oscillate about this average distance. As the temperature rises and the thermal agitation increases, the molecules elbow themselves farther and farther apart until at last they are just at the boundaries of their spheres of effective attraction. A little excess of thermal agitation in some part of the liquid will then enable a group of molecules to push their surrounding neighbors quite out of the way, and form a bubble of vapor. This occurring repeatedly and at many places, the liquid boils. Again the energy for pushing the molecules apart is drawn from the heat supplied, and is represented by the latent heat. There is here no distinction between crystalline and amorphous bodies, for all liquids are structureless. All have sharp boiling points, and absorb a large latent heat in boiling.

Finally, in the gaseous state, the molecules released from all effective restraint fly about independently, and push and jostle one another apart indefinitely, or until they are stopped by the walls of a retaining vessel, or in the case of an atmosphere by the gravitational attraction of the planet. As early as 1738, the mathematician Daniel Bernoulli suggested that the indefinite expansion of a gas was thus due to its flying molecules, and not to a repulsive force as Newton had supposed. He was even able to show mathematically that on this assumption a pressure would be produced, by the impacts of the molecules on the walls of a retaining vessel, that obeyed Boyle's law. In 1857 Clausius further showed that if the average kinetic energy of the flying molecules is proportional to the absolute temperature of the gas, Gay-Lussac's law would also be obeyed. These two propositions form the basis of the modern kinetic theory of gases.

At a given moment, of course, the molecules will not all be flying with the same speed. Some will be flying faster, others more slowly than the average. It is the *average* velocity that determines the temperature. Similarly in a liquid or in a solid, the molecular heat agitation will vary from place to place about a mean value. These fluctuations were shown by Maxwell in 1859 to follow the laws of probability. This is the basic theorem of the modern statistical mechanics, which, combined with the quantum theory, is to-day largely displacing the mechanics of Newton, at least so far as atomic and subatomic physics are concerned, and leading to a "Principle of Uncertainty" and other startling things.

VII

EVAPORATION

SOMEWHAT akin to the vaporization of a liquid that takes place at the boiling point, is the slower process of evaporation that occurs at all temperatures. The rate of evaporation increases with the temperature, and depends upon how near the temperature is to the boiling point of the liquid. Hence liquids of low boiling point are more volatile at ordinary temperatures than those of high boiling points. But whereas boiling is a process that involves the whole liquid, evaporation takes place only at the exposed surface, hence increases when the latter is enlarged.

Above the surface of a liquid, there is always a certain amount of its vapor. The presence of this vapor is a hindrance to further evaporation. Its removal promotes the process. Therefore, when a pan of water is set out in a breeze, it dries up faster than when set out where the air is stagnant. If, instead, a liquid is shut up in a bottle tightly corked, so that none of the vapor can escape, then the space above the liquid soon becomes so charged with vapor that it can hold no more. Evaporation then ceases, and the vapor is said to be *saturated*.

Heat is required for evaporation just as it is for boiling. But in this case the heat is drawn from the liquid itself, which is thereby cooled. The faster the evaporation, the greater the cooling. It is therefore greater for the more volatile liquids, such as alcohol, gasoline, or ether, as one may roughly verify by sticking his finger successively in these liquids, and then holding it up to the air.

The cooling produced by evaporation is used in many ways. Nature makes use of it to cool our bodies, by bathing them with perspiration. We promote the process by fanning ourselves, thus driving the vapor away. Similarly we cool our soup by blowing on it—lightly, so that no one will notice. The

Mexicans cool their drinking water by keeping it in porous vessels. The water seeps through to the outside and evaporates. The surgeon uses a very volatile liquid to freeze a portion of the skin that requires a minor operation.

These phenomena of evaporation are readily explained by the kinetic theory. Suppose that a molecule on the surface of a liquid happens to be considerably more agitated than the average. Its super-agitation may become sufficient to enable it to break away from its neighbors, and shoot out of the liquid. Every second, in fact, there will be a certain proportion of the surface molecules that reach this degree of excess agitation, and escape. The proportion increases as the temperature rises. Hence the rate of evaporation increases with the temperature. When the boiling point is reached, the *average* agitation becomes equal to that required for escape, so that the molecules no longer escape only one by one and only from the surface, but in bunches and within the body of the liquid, wherever a slight excess of agitation enables them to push their fellows aside and form a bubble.

Now, since it is only the more lively molecules that can jump out of the liquid, the average kinetic energy of those that remain behind is reduced. Hence the liquid is cooled. And since the escaping molecules carry away more than their fair share of the kinetic energy, the vapor is warmer than the surroundings. But if these livelier molecules hang around, some of them are likely to shoot back into the liquid, and in fact do so. This of course reduces the net rate of evaporation. Hence if the vapor is removed by a breeze, by an air pump, or otherwise, the net rate of evaporation and the cooling of the liquid are increased. In fact, a liquid may be made to freeze itself by this means. If, instead, the liquid is put into a sealed container so that the vapor cannot escape, then, as evaporation proceeds, the vapor becomes continually denser; increasing numbers of molecules shoot back into the liquid until the number shooting back are equal to those shooting out. The liquid is then said to be in *equilibrium* with its vapor. No further change then takes place in the proportion of liquid to vapor, and both assume the same temperature.

At low temperatures evaporation becomes exceedingly slow. Nevertheless it goes on. There will always be now and then a molecule that attains the critical escape velocity, just as in a

sufficiently large group of men there will always be an outstanding individual, no matter how dull the rest.   Since in a
drop of water, there are at least a billion times a billion molecules, the chances are good that a few will be highly exceptional.

The atmospheres of planets "evaporate" in a very similar
fashion.   Every now and then in the upper layers, a few molecules acquire a high enough outwardly directed velocity to
escape from the gravitational attraction of the planet, and
wander off into space.   The "escape" velocity for the earth is
seven miles per second, for the moon one and one-half, for
Mercury two and one-half, for the sun 380 miles per second.   It
depends upon the size and density of the body.   Also, the
heavier the gas, the more sluggish its molecules.   Hence the
earth retains oxygen and nitrogen, but not helium or hydrogen.
The moon and Mercury retain no atmospheres at all, whereas
the sun and stars retain all the gases including great quantities
of helium and hydrogen.   But however strong the gravitational
attraction and however heavy the gases, *some* evaporation goes
on, so that the stars and planets would eventually lose all their
atmospheres, if the supplies of gas were not constantly renewed
from the body of the planet or star.

How far does the earth's atmosphere extend?   That is a
difficult question to answer.   As we ascend, the air becomes
rarer and rarer, so that the question becomes—when is a gas
not a gas?   When do the molecules become so scattered that
they must be considered as individual wanderers, rather than as
members of a community?

Meteor trails have been observed at a height of one hundred
miles, so that at this height the molecules, few as they may be,
are still fairly uniformly distributed.   But somewhere above this
region we enter another, which is occupied only by molecules
that have shot out from the region below, but with insufficient
velocity to escape, and which therefore fall back again.   This
region is full of cascading molecules which shoot out in all
directions and fall back again, like the stars of a Roman candle.
Still higher we come to a region into which some molecules have
shot with a velocity so great that they no longer fall back, but
yet the velocity is not sufficient to carry them away entirely.
Hence they execute elliptical orbits about the earth, much like
comets about the sun.   Still beyond we shall encounter here
and there a molecule that has shot out with parabolic velocity

or more, and which is leaving the earth forever. One cannot say how high the atmosphere extends, or if it ever really ends.

Even solids evaporate, though usually, of course, with extreme slowness. Nevertheless, in some cases, and particularly near their melting points, the evaporation of solids becomes quite appreciable and important. Thus naphthalene moth balls, which melt at 176 degrees Fahrenheit, considerably under the boiling point of water, reduce in a few months' time to the size of peas, and eventually disappear altogether. The filaments of tungsten lamps, which run at temperatures as high as 5000 degrees Fahrenheit, while the melting point of tungsten is 5600 degrees, evaporate, and the vapor condenses on the cooler walls of the bulb. This causes the blackening of the bulb. The whole arrangement is virtually a distilling apparatus, which causes the lamp to dim with age, and ultimately to burn out.

Everything, then, evaporates. All liquids and solids, whatever their temperatures, are slowly being converted into vapor. The greater part of the universe is already gaseous. All the stars are gaseous, and enormous masses of gas, as bright or as dark nebulae, are floating about in space. Only on the earth and on other small cold bodies do we find liquids and solids predominating. We live amid exceptional circumstances.

This universal evaporation was once made the basis of an argument, by the astrophysicist Zöllner, that the universe could not possibly have had an infinite past. In that case, he said, everything would have long since already evaporated.

VIII

EFFECTS OF PRESSURE

In the preceding chapters we have spoken of the boiling points of liquids, as though each liquid had but one, and that one were always the same. That is not true. The boiling points vary with the pressure. In order for a bubble of vapor to form in the body of a liquid, it must push back the surrounding liquid that presses in upon it from all sides. This pressure, a short distance below the surface, is substantially the same as that of the atmosphere (or of whatever other gas or vapor may be present) upon the surface of the liquid. If this pressure is increased, a greater expansive force, a greater thermal agitation, a higher temperature, will be required to form a bubble. Increase of pressure, therefore, raises the boiling point, and, conversely, decrease of pressure lowers it. When we speak of *the* boiling point of a liquid, we mean its boiling point at a pressure of one standard atmosphere, which is 14.7 pounds per square inch, and corresponds to a barometer height of 76 centimeters or 29.92 inches.

If we stop up the spout of a kettle and clamp down the lid, the pressure within will rise, and so will the temperature. This is the principle of the steam boiler and of the pressure cooker. By allowing the pressure to increase to a point regulated by the safety valve, the temperature of the water can be raised considerably above its ordinary boiling point. This means a more efficient engine in the one case, and more effective cooking in the other. Conversely, if we attach a hose to the spout, connect it with an air pump, and draw off the vapor, the water will boil at a lower temperature. When one ascends a high mountain, the air pressure diminishes, and boiling takes place at a lower temperature. For this reason it is difficult to make good coffee, or to cook by boiling, unless a pressure cooker is used.

The lowering of the boiling point by reduced pressure can

easily be shown by the simple apparatus of Figure 22. The inverted flask contains water just under 212 degrees Fahrenheit. By squeezing a sponge full of cold water over the flask, some of the vapor within is caused to condense on the glass. This reduces the pressure, and the water immediately boils.

A more satisfactory and complete demonstration, however, can be made with an air pump. By rapidly exhausting the air and vapor above a liquid, the boiling temperature can be lowered any amount we please, even down to the freezing point.

FIG. 22—BOILING AT REDUCED PRESSURE

This principle is applied in the vacuum pan, used in the manufacture of sugar and in other industries. The juices which have been squeezed out of the sugar cane must be boiled down to remove the greater part of the water. But sugar is very easily burnt. Hence the boiling temperature is lowered by reducing the pressure by means of a pump. One might expect in this way to save some of the fuel required for the boiling, but the working of the pump consumes more than is saved.

A knowledge of the temperatures at which water boils under different pressures is of importance in two great fields—steam engineering and meteorology. Accordingly in 1847 Regnault, at the instance of the French Government, began a long series of accurate measurements covering a large range of pressures. His results, extended by later investigations, are shown graphically by the curve in Figure 23, which is called the *steam line*. The boiling points of other liquids give similar curves, which we shall also call *steam lines* for short.

On account of the great range of pressures involved, it is impossible to get the whole of this curve within the compass of a single conveniently sized diagram, if it is drawn to a uniform scale in the usual way. We have therefore plotted horizontally

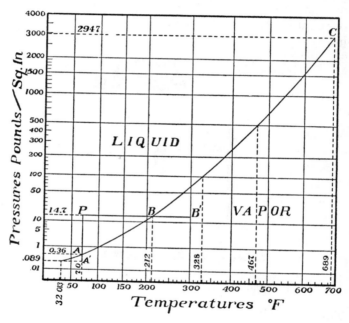

FIG. 23—THE REGNAULT STEAM LINE

the cube roots of the temperatures, and, vertically, the fourth roots of the pressures. By this device the diagram is compressed in the region of the higher pressures and temperatures, and stretched out in that of the lower ones. This is desirable because the lower part of the curve is the more important, for pressures above four hundred pounds are seldom used. If the pressures had been plotted to a uniform scale, each pound being represented, say, by the height at which one pound stands on this diagram, the four hundred-pound level would have stood at a height of about thirteen feet, while the rest of the diagram would have extended to over one hundred feet. This, to say the least, would be very inconvenient. It was not so necessary to compress the temperature scale, but it has been done to prevent undue distortion of the curve, and to make this diagram correspond to that of Figure 36, on page 109, of which it is a part. The real temperatures and pressures, not these roots, have been marked on the scales. Hence they can be read in the usual way.

From the curve we can read the boiling points corresponding to various pressures. For example, at a pressure of one atmosphere, 14.7 pounds, as read on the vertical scale, we find the boiling point at B on the curve to be 212 degrees, which number is read on the bottom scale. Similarly we may find from the curve, that in a steam boiler at 100 pounds pressure, the temperature will be 328 degrees, and at 500 pounds, 467 degrees, etc. We must here remark, that the pressures shown by a steam gauge are pressures above the atmosphere. To get the total or *absolute* pressures, which we here use, that of the atmosphere must be added to the gauge readings. We should also say that for exact work, readings from a curve drawn to a small scale, such as this one, are much too inaccurate. For such work, a *steam table* is used, where all these results and many others useful to engineers are tabulated.

Conversely, we can find from the curve the pressures at which water boils for given temperatures. For example, if we wish to boil water at room temperature, 70 degrees, we shall find that we must reduce the pressure to 0.36 pounds. If we wish to boil it at the freezing point, we must reduce the pressure to 0.089 pounds, corresponding to a barometer height of 0.18 inches, which is a pretty good vacuum.

The highest boiling point given on the curve is 689 degrees, which is attained at a pressure of 2947 pounds per square inch. Here the curve abruptly ends, for the reason that above this temperature water cannot exist in the liquid form, no matter what the pressure is. This is called the *critical temperature*, of which more later.

For a given pressure, the temperature at which water boils is the same as that at which steam condenses. It is therefore indifferent whether we say that the steam line gives the boiling points of water, or the condensing points of steam. They are the same thing. At every point of the curve a *mixture* of both states is present. The water is at the boiling point, the steam

is at the condensing point.   If heat is added to the mixture, the
water boils.   If heat is abstracted, the steam condenses.   Hence
the two states are in equilibrium, and the steam is saturated.
The curve is the *equilibrium line* for a mixture of water and
steam.   At all points of the diagram to the left and above the
curve, the substance is wholly liquid.   At all points to the
right and below, the substance is wholly vapor or gas.   The
curve is hence the boundary line between these two states, and
wherever we cross it, we pass from the one state to the other.

Hence a vapor in contact with its liquid is always saturated.
It is impossible to raise its temperature above the boiling point
corresponding to the pressure, for the addition of heat merely
causes more water to boil off, and the temperature remains
unaltered.   Conversely, it is impossible to cool the vapor below
this temperature, for the abstraction of heat merely causes the
vapor to condense, again at constant temperature.   At least this
is ordinarily the case.   We shall note some exceptions later.
But if a vapor is separated from its liquid (drawn into a separate
container), its temperature can be raised above the boiling
point corresponding to the pressure by any amount we please.
The vapor is then said to be *superheated*.   This is the condition
of the substance *below* the steam line.

If a small amount of liquid is shut up in a sealed container,
so that the total volume is kept constant, the liquid, as we have
seen, evaporates until the vapor in the space above it becomes
saturated; that is, until the pressure of the vapor is that for
which, at the prevailing temperature, the vapor is just about to
condense.   This is also the *pressure* at which the liquid would
boil at the prevailing temperature.   Hence the vapor is at the
condensing point, and the liquid is at the boiling point.   We
are on the steam line.   If heat is added to the mixture, its
temperature rises, more water evaporates, and the *vapor pressure* rises until it is again at the saturation pressure corresponding to the higher temperature.   We are still on the steam line.
We have simply moved up a bit.   Conversely, if heat is removed,
the temperature falls, vapor condenses, and the vapor pressure
diminishes until it is again at the saturation value corresponding
to the lower temperature.   We move down the steam line.

What we have described is a species of boiling and condensing at *constant volume*, whereas these processes are commonly
carried out at *constant pressure*.   Thus, if we cork up some water

in a bottle at 70 degrees Fahrenheit, the water will evaporate until the vapor pressure reaches 0.36 pounds, and then evaporation stops. This is the saturation pressure for that temperature —and the liquid is at the boiling point for that pressure. If we raise the temperature to 100 degrees, the vapor pressure rises to 0.95 pounds; at 212 degrees it reaches one atmosphere or 14.7 pounds; at higher temperatures, it reaches still higher pressures, always as indicated by the steam line at the corresponding temperatures. We have now virtually a steam boiler, and this in fact is precisely what happens in such a boiler while we are "getting up steam," for this is also a constant volume operation. The pressure and temperature conjointly rise, keeping always to the steam line, until the opening of the safety valve puts a stop to the pressure rise, and consequently also to the temperature rise, and boiling then proceeds in the usual way at constant pressure and at constant temperature.

If we have the proper proportions of liquid to vapor in a sealed container, we can in this way run all the way up and down the steam line, and even run off the end. What happens then will be described later. If there is too little liquid, all of it may boil off at some point, and we then branch off into the vapor region. If there is too much liquid, all the vapor may condense prematurely, and we then run off into the liquid region.

In the ordinary processes of boiling and condensing, the pressure is *maintained* constant by external circumstances. The temperature then automatically remains constant, and cannot be altered. Conversely, if the *temperature* is maintained constant by external means, boiling and condensing then take place at constant pressure, and the pressure cannot be altered. Suppose, for example, that a mixture of liquid and vapor is contained in a cylinder having a movable piston, and that the whole is immersed in a large water bath by which the temperature is kept constant. We endeavor to compress the vapor by pushing in the piston. We shall not succeed, at least if we go slowly. The pressure remains unchanged, the vapor merely condensing. Ordinarily the compression of a gas heats it. But in this case the heat of compression is at once absorbed by the water bath, so that the temperature is not changed. What we do in effect is to squeeze the latent heat of the substance out into the surroundings. Conversely, if we endeavor to reduce the pressure by withdrawing the piston, we shall not succeed. The liquid at once boils, and keeps the space filled with its vapor at the saturation pressure. In this case we suck up, so to speak, the latent heat required for the boiling from the surroundings.

A form of this experiment can readily be carried out with the apparatus shown in Figure 24. This consists of two barometers *a* and *b*, which are prepared in the following way. A glass tube, something over a yard in length and closed at one end, is filled with mercury, and then while the open end is stopped with the finger, the tube is inverted, and this end dipped into a cistern of mercury. The finger is removed, and the mercury in the tube then descends to a height of about thirty inches above the level in the cistern. This represents the atmospheric pressure. In descending, the mercury leaves above it a completely empty space—the Torricellian vacuum. The tube at *a* is left in this condition. Into the one at *b* is introduced a small amount of ether. This is done by means of the little syringe with the crooked spout shown at *C*. The spout is inserted under the open end of the barometer, and some of the contained ether is squeezed out. The ether ascends the tube in bubbles till it reaches the top. The first few drops evaporate immediately, and the pressure of the vapor thus formed depresses the mercury column. The depression increases as more ether is introduced. But at last it stops, and some of the ether remains in liquid form on top of the mercury. More ether then produces no further depression, except by the very light weight of the additional liquid. We have now reached the *maximum* or saturation pressure of the ether vapor. The condition of the apparatus at the end of these operations is shown at *A* in the figure.

We could have used water for this experiment. But the saturation pressure of water vapor at 70 degrees is only a third of a pound, and this would have depressed the mercury column by only two-thirds of an inch. Ether is much more volatile. Its saturation pressure at 70 degrees is 8.4 pounds, and this depresses the mercury column 17 inches. The effect is conspicuous and surprising.

The mercury cistern has a deep well in the center into which the tube *b* can be lowered. As we do this, we shall observe that the mercury column does not sink with the tube, but remains at a fixed height, so that the space above it in the tube diminishes. At the same time the quantity of liquid ether in the tube increases. This stage of affairs is shown at *B*. Some of the vapor has condensed. But the constant height of the mercury column shows that the vapor pressure has remained unchanged. It is still at the saturation value of 8.4 pounds, and still depresses the mercury column by 17 inches. The

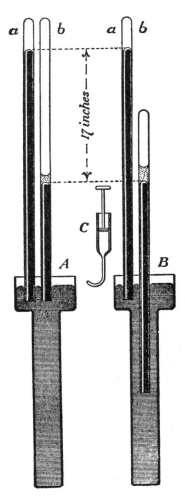

FIG. 24—DEEP CISTERN BAROMETER EXPERIMENT

tube must of course be lowered slowly to give time for the latent heat released by the condensation to be dissipated to the surroundings.

If the tube is further lowered, the condensation continues, still without change of pressure, until the ether is completely liquefied. Further lowering of the tube then pushes the mercury down, for the top of the tube is now against a plug of liquid, which is well-nigh incompressible.

By raising the tube, all these operations are reversed. The mercury rises with the tube until the ether begins to boil. From then on it remains at a constant height until all of the ether has vaporized. If it is possible to raise the tube still further without pulling it out of the cistern, the mercury column will again rise, but more slowly now than the tube, showing that as the volume of the now superheated vapor is increased, its pressure diminishes, just as occurs with any gas.

The ordinary process of boiling, as when we put a kettle of water on the stove, is represented on the Regnault diagram (page 66) by the constant pressure line PBB'. We start at P with the water at 70 degrees, and one atmosphere pressure. As we add heat, the temperature rises, and we move along PB, until at B we reach 212 degrees and boiling begins. Here, there is a long pause in our progress along the line, for the temperature does not again rise until all the water has boiled off. Then the steam remaining in the kettle is superheated along the line BB', and could have been continued indefinitely. If now we allow the kettle to cool, we retrace our steps. The steam cools along $B'B$. At B the saturation temperature is reached, and condensation begins. Again a long pause until this process is completed, and only water is present. The latter then cools along BP, until at P we are back at the starting point.

We have now less water than before because most of the steam has escaped. But suppose we had heated the water at the bottom of a very tall cylinder provided with a frictionless piston, whose weight was such that it maintained a constant pressure of 14.7 pounds on the contents of the cylinder. The steam resulting from the boiling would then simply push the piston up, and all of it would be saved. The process would then be completely reversible. In this case we must note that the piston and the sides of the cylinder must be non-conductors of heat, for otherwise the steam will condense from loss of heat.

Also the cylinder must be very tall, for the volume of the steam is sixteen hundred times that of the water from which it is derived. If we started with an inch of water on the bottom of the cylinder, the latter would have to be 133 feet high to accommodate the resulting steam. This large expansion and the similar contraction during condensation take place during the pause at *B*. They do not show on the Regnault diagram, because this being a *pt* plot, shows only relations of pressure and temperature. To show also the volume changes would require another dimension.

A liquid, we have seen, can also be brought to boil without changing its temperature, by simply reducing the pressure, as by means of a vacuum pump. Since the function of the pump is merely to enlarge the volume, we may imagine this process also carried out in a very tall cylinder provided with a piston which we gradually raise. We thus produce the required enlargement of volume by a single long stroke, instead of by a succession of short ones as with the pump. The process is essentially the same as that of our deep cistern barometer. Let us start again with water at 70 degrees, and the piston pressing directly upon it with a pressure of 14.7 pounds per square inch. As we raise the piston the pressure diminishes, and we must now suppose the cylinder such a good conductor of heat, that the prompt inflow of heat from the surroundings keeps the temperature constant. We thus descend the constant temperature line *PAA'* on the Regnault diagram. At *A* we reach the boiling *pressure* corresponding to 70 degrees. Here there is a *very* long pause in the descent, for although the piston is being continually withdrawn, the liquid is vaporizing, and the pressure does not drop. The increase in volume at this temperature is much greater than at 212 degrees. It is no less than fifty-four thousand times. Hence, if we start with an inch of water, we shall require a tube forty-five hundred feet high to accommodate the vapor. This tremendous volume is due to the low pressure, 0.36 pounds. When at last the liquid is completely vaporized, further withdrawal of the piston reduces the pressure, and we continue the descent along *AA'*. Despite the moderate temperature, the steam is now superheated, for at *A'* the pressure is 0.089 pounds, which corresponds to a boiling point of 32 degrees. The steam at 70 degrees is therefore above the vaporization temperature corresponding to its pressure, and so by our definition above is superheated. Superheated steam is therefore not necessarily excessively hot, as its name would suggest, although the kind that the engineer deals with usually is. But the meteorologist often has to do with superheated steam that is extremely cold, that is even below the freezing point.

We can now reverse our whole experiment by pushing the piston in, condensing the vapor, and eventually returning to the starting point. During the expansion process, heat had to be absorbed from the surroundings to keep up the temperature, and to provide the latent heat of vaporization. During the contraction, heat had to be rejected to the surroundings to keep the temperature down, and to remove the latent heat.

The freezing points are also affected by pressure, though to a far less extent than the boiling points. Also we find opposite effects depending upon whether the substance expands or contracts on solidifying. If it expands, pressure will hinder the process; if it contracts, pressure will aid it. Hence we may reasonably expect that when the pressure is increased, bodies that expand will require a greater cold, those that contract, a lesser cold, to solidify them. These are indeed the facts. Increase of pressure lowers the freezing point of a substance like water which expands, and raises that of one like paraffin which contracts on solidifying. The effect, however, is extremely small. An increase in pressure of one atmosphere lowers the freezing point of water only 0.013 Fahrenheit degrees, and raises that of paraffin 0.036 degrees. The effect is very nearly proportional to the pressure, so that at seventy-eight atmospheres, or 1170 pounds per square inch, the freezing point of water is lowered one degree. The effect depends also upon the magnitude to the volume change on solidification, so that for nearly all other substances it is smaller than for water or paraffin.

It was at one time thought that the force of expansion when water froze was irresistible. The idea seems to have been that if the temperature is below freezing the water simply *must* freeze, *must* expand, and therefore *must* overcome any resistance that opposed it. Indeed, very powerful effects can be produced. Great weights can be lifted by the expansion of water on freezing. Huygens succeeded in bursting a cannon by freezing water contained in it. But there are no irresistible forces in nature. This giant though strong is not of unlimited strength. If the temperature is 31 degrees Fahrenheit, a part of the water in a confined space will freeze until the expansion has produced a pressure of 1170 pounds. If the container is strong enough to withstand that, the rest of the water simply will not freeze, and there will be no bursting. If the temperature is 30 degrees, the freezing will stop at 2340 pounds, and so on. At each temperature there is a definite limit to the expansive force. Besides, modern investigation has shown that at −8 degrees Fahrenheit, at which a pressure of 31,500 pounds per square inch is reached, the freezing point ceases to be lowered by increased pressure. Thereafter it rises, and the water, instead of expanding, contracts on freezing, as will be more fully explained in the next chapter. 31,500 pounds is hence the utmost pressure that can be obtained by freezing water in a confined space. Containers have been made that can withstand that pressure and a good deal more, so the frost giant has been successfully chained.

The lowering of the freezing point by pressure means that ice can be melted by pressure. When the pressure is relieved, the water freezes again. This phenomenon is called *regelation*. It has many important consequences.

Every skater knows that when the weather is very cold, and the ice is consequently dry, the skating is not good. The ice does not seem to be even slippery. In fact dry ice is not especially slippery. It is the film of water on wet ice, which, acting as a lubricant, makes it slippery. Now, as Joly pointed out in 1899, the pressure under a skate is very great. On account of the curve of the blade and the hollow grinding, the area on which a skate rests is very small, one-fiftieth of a square inch or thereabouts. With a skater weighing one hundred and forty pounds, a pressure of seven thousand pounds per square inch is produced. This is sufficient to depress the freezing point 6.2 degrees, so that if the temperature of the ice is not under 26 degrees Fahrenheit, it will be melted by this pressure. The skater therefore glides along on a film of water, which, if the temperature is below 32 degrees, recongeals after he has passed.

It is this regelation that enables a glacier to descend a sinuous valley. Wherever an obstruction is met, a great pressure is produced, which melts the ice. But the water passes around the obstruction and freezes on the other side of it, where the

pressure is relieved.   Hence it appears as though the ice itself
had bent around or over the obstacle like a flexible thing.   The
process can be strikingly demonstrated in the following way.
A block of ice is supported between two tables, and a loop of
thin copper wire is passed around the middle of the block, and a
rather heavy weight is attached to it.   The pressure of the wire
melts the ice immediately beneath it, but the resulting water
passes around to the top of the wire, and refreezes.   Consequently the wire gradually eats its way through the ice, and will
eventually pass completely through it and the weight fall to the
floor.   But the block of ice remains undivided and as solid as
ever.   By pressing two pieces of ice together they can be made
to freeze together, and, as every one knows, the glacier itself is
formed by the compacting of great masses of snow in the upper
part of the valley.

IX

THE TRIPLE POINT AND
HIGH PRESSURES

SINCE decrease in pressure raises the freezing point and lowers the boiling point of water, at some reduced pressure the two should meet. This meeting can in fact be brought about by the following experiment devised by Leslie. A small dish of water and a large dish of strong sulphuric acid are placed under the receiver of an air pump. The air is rapidly exhausted, and the water vapor is largely absorbed by the sulphuric acid, which has a strong affinity for it. The rapid evaporation lowers the temperature of the water faster than the boiling point descends, so that the water does not boil until the rising freezing point is met. Then it *boils and freezes at the same time.* Bubbles of steam rise from the interior of the liquid—very cold steam, for they freeze immediately on reaching the surface. The process continues until the whole is a solid block of ice, which can then be removed from the receiver. Since the utmost that the pressure can be reduced is from one atmosphere to zero pressure, and this we have seen would raise the freezing point only to 32.013 degrees Fahrenheit, it is obvious that the meeting point must lie between this temperature and 32 degrees. It lies, in fact, at 32.01289 degrees, and at a pressure of 0.089 pounds per square inch. The ice is hence about as cold as that produced in the ordinary way.

This experiment can also be carried out with a substance whose freezing point is lowered by diminished pressure, because the descent of the freezing point is so much slower than that of the boiling point, that the latter soon overtakes it.

This point where a substance boils and freezes at the same time is called the *triple point*, because solid, liquid, and vapor there exist together. This point for water is the beginning of

the Regnault steam line of Figure 23, and represents the lowest
pressure at which water can exist.   If we plot the pressures and
corresponding temperatures at which water freezes, we get
another line on this diagram, the *ice line*, which also starts at
the triple point, as shown in Figure 25.   This line leans to
the left, in the direction of decreasing temperature, because
increase of pressure lowers the freezing point.   For a substance
which contracts on solidifying, this line leans to the right, as
shown in Figure 34, Chapter XI.   The steam line, it will be
remembered, divides the Regnault diagram into a liquid and a
vapor region.   The ice line further subdivides the upper part of
the diagram into a liquid and a solid region, and the line represents a *mixture* of liquid and solid in equilibrium.

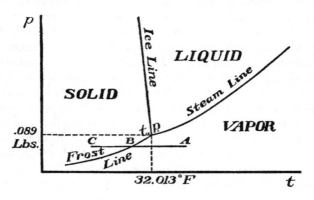

FIG. 25—THE TRIPLE POINT

Yet a third line starts at the triple point.   This divides the
solid from the vapor region.   It is called the *hoar frost* line.
In crossing this line, the vapor transforms directly into the solid
(the process of sublimation) or the solid into the vapor.   This
can only occur at pressures and temperatures below those of the
triple point, as along the constant pressure line *CBA* in the
figure.   At *B* direct vaporization of the solid occurs if heat is
added, and sublimation, if heat is abstracted.   On the line
there is always a mixture of solid and vapor in equilibrium.
The same sort of diagram applies to substances other than
water, but we shall retain the names ice and frost lines for
short.

In winter when the air in the house is dry, which means that the pressure of the water vapor contained in it is very low, and the windowpanes are below 32 degrees, this vapor deposits on the panes directly as frost. It does not condense first to water and then freeze. In that case clear ice would be produced. The beautiful frost patterns result only from the direct deposition of the vapor in solid form. Similarly, snowflakes are crystallized vapor, whereas hailstones are simply frozen raindrops.

The direct passage of ice to vapor and the reverse are of rather rare occurrence, because the triple point of the substance is so low. But for many substances this point lies much higher, so that the phenomena occur at ordinary pressures and temperatures, and even much above. Sulphur and iodine are examples. Frozen carbon dioxide melts directly to vapor, for which reason it is popularly called dry ice. The steam we see rising from it is not its vapor, but is the water vapor of the air, which has been condensed by the cold but invisible carbon dioxide vapor. If the dry ice is placed in a pan of water, this apparent steaming is much increased.

If the triple point pressure of a substance is above one atmosphere, the substance can exist in the liquid form only under increased pressure, and this no matter how much the temperature is lowered. This is the reason why we never see iodine, carbon dioxide, or other such substances in the liquid form, except perhaps in a glass vessel under pressure.

The most remarkable substance in this regard, however, is carbon. In the electric arc it vaporizes directly. In the bright parts of flames its vapor precipitates at once into solid particles, as is evidenced by the fact that they will deposit on a colder body as soot. No one has ever seen carbon in the liquid state. Evidently both the pressure and the temperature of its triple point are extremely high. We must imagine tp in Figure 25 shifted a long way upward and to the right, so that nearly all of the diagram is solid and vapor, and only a little upper corner is liquid. If diamonds are crystallized from liquid carbon, that would account for their rarity. Moissan, proceeding on this theory, cooled white-hot iron containing small specks of carbon very suddenly. The contraction of the iron produced a great pressure on these particles, and he obtained in this way microscopic diamonds. We can produce in our laboratories very high temperatures, or very high pressures, but not both together.   As yet, nature alone can do this, in the depths of the planets and stars.

We have mentioned that solids show minor changes of state having all the characteristics of those major changes from solid to liquid and liquid to gas, the chief of which is a change in volume at constant temperature and pressure, accompanied by the absorption or rejection of a latent heat.   In the year 1900, Tamman[1] carried out a remarkable series of experiments, the primary objects of which were to determine whether the depression of the freezing point of water continued indefinitely

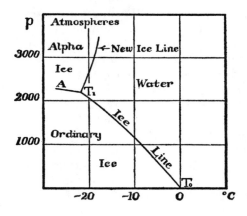

FIG. 26—TAMMAN'S ALPHA ICE

as the pressure increased, and whether the rate was constant. For this purpose he carried the pressure up to the hitherto unheard-of value of 3500 atmospheres, or 52,000 pounds per square inch.   He found that the rate of depression increased as the pressure rose, so that the ice line of Figure 25 *curves* to the left.   But at − 8 degrees Fahrenheit and 2200 atmospheres, an abrupt and remarkable change took place.   The ice suddenly contracted by about 20 per cent so that its volume was now *less* than that of water, and with further increase in pressure, the melting point now rose, as we have mentioned in the previous chapter.   From this point on, the ice line leans to the right, as shown in Figure 26, which is Tamman's diagram.   In short, a change of state had taken place.   A new form of the solid had appeared.   Tamman called it alpha ice.   It is obvious also

[1] Tamman, *Annalen der Physik*, Vol. II, p. 5.

that the point on the ice line where this change took place is a new triple point, for here the old and the new ice and water all three meet. Tamman, indeed, was able to follow the equilibrium line that separates the two forms of ice for a short distance, as shown at T_1A in the figure. The ordinary ice line T_0T_1 leans so much more to the left in this diagram than in Figure 25, because of the small scale on which the pressures are plotted. The ordinary triple point T_0 on this scale is practically at zero pressure.

In these experiments, the ice or water was compressed in a strong steel cylinder, and so of course could not be seen. Only the sudden change in volume at constant temperature and pressure announced a change of state. But Tamman was anxious to get a look at the new ice if possible. So he cooled it, still at high pressure, to the temperature of liquid air, released the pressure, and got it out of the cylinder. It looked just like any other ice, but was extremely cold and dry. As it warmed up, it gradually melted into ordinary ice, with a large increase in volume.

In later experiments Tamman discovered a third kind of ice, and thought he had found indications of a fourth, but the latter has not been confirmed.

In 1911 P. W. Bridgman[2] of Harvard University, having invented a new kind of packing, was able to carry the pressure up to 40,000 atmospheres or 600,000 pounds per square inch, nearly twelve times as far as Tamman. This is the pressure that would be produced by a pile of bricks 100 miles high, or that would be found 80 miles beneath the surface of the earth. At these stupendous pressures some profound changes occur in the properties of matter. Bridgman found that steel became plastic and could be made to flow like pitch. Soft rubber, instead, became hard and brittle like glass. The rubber washer would crack and the softened steel flow into the cracks.

With this apparatus Bridgman extended Tamman's experiments, discovering five different kinds of ice, which he numbered I, II, III, V, and VI. Number IV was reserved for Tamman's fourth variety, but this was not found. Number I is ordinary ice, III is Tamman's alpha ice. Of 11 possible transition lines, 10 were found; of 6 triple points, 5 were found. All of these forms of ice except Number I are of less volume than

[2] Bridgman, *Proc. Amer. Acad. Arts and Sciences*, Vol. 47, p. 441.

water.   Figure 27 is Bridgman's diagram.   The temperatures
are given in centigrade degrees.   On the Fahrenheit scale the
range is from −112 to 176 degrees.   The pressures are in
kilograms per square centimeter, or what are called metric
atmospheres, which are ninety-seven hundredths of ordinary
atmospheres, hence practically the same.   Ice Number VI
seems to be the final form.   No further change was found at
the highest pressures and temperatures used.   And in this form
the freezing point rises continually with the temperature.

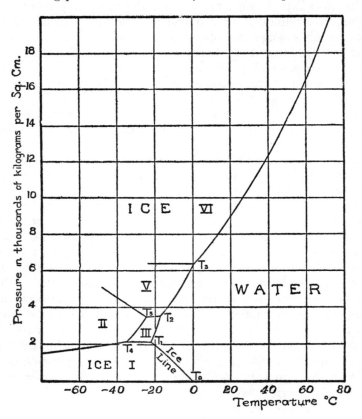

FIG. 27—BRIDGMAN'S FIVE FORMS OF ICE

At the time of the gold rush to the Klondike, many were the
tales told of the cold encountered.   Perhaps one of the tallest
was that told by a man who said that one day he set a bucket of

hot water out on his porch to cool, and the water froze so quickly that the ice was still hot. While this may be an exaggeration, Bridgman's ice Number VI at 160 degrees Fahrenheit is distinctly hot ice. But we can only have it at 300,000 pounds pressure, so that we shall probably never have a chance to burn our hands on it. If we could only produce pressure enough, it appears that we could have ice at any temperature we please, even above the ordinary boiling point of water.

Bridgman also experimented with frozen mercury and with many other substances at these extraordinarily high pressures. He found that many of them exhibited similar changes of state within the solid state.

X

SOLUTIONS

It had long been known that the boiling point of a liquid is raised when a substance is dissolved in it. But in 1822, Faraday, while measuring the boiling points of various aqueous solutions, made the curious discovery that when he lifted his thermometer so that the bulb was in the steam just above the boiling fluid, it always dropped to 212 degrees Fahrenheit. Hence it appeared that, however elevated the boiling point of the solution might be, the steam remained at 212 degrees, the boiling point of the pure solvent.

Gay-Lussac, on hearing this, objected that the temperature of the steam could not possibly be appreciably lower than that of the liquid from which it had just been disengaged, and concluded that either Faraday's thermometer was wrong, or that he had not observed with sufficient care.

But Faraday's observations and his thermometer were correct. The latter does read, and rightly reads, 212 degrees in the steam. Yet Gay-Lussac's objection was entirely justified. This is not the temperature of the steam. Now, how can a thermometer be right if it registers the wrong temperature? The answer is that the bulb is wet, and the water that has condensed upon it is pure water—it is distilled water, for the solids remain behind when a solution is boiled. If the thermometer falls below 212 degrees, water condenses upon it and, giving up its latent heat, warms the thermometer. If it rises above 212 degrees, this water boils off, and being pure water it boils at 212 degrees and, by abstracting its latent heat from the thermometer, cools the latter. When we put a kettle of water on the stove to boil, we surround it by flames that may be a thousand degrees hotter than the water. Yet we cannot raise the temperature of the water above 212 degrees. And so the hot steam surrounding the thermometer is powerless to raise its temperature above the boiling point of the pure water that covers it. Wet the bulb of a thermometer exposed to the air. Its temperature immediately drops, because of the evaporation of the water. And if the supply of water is kept up, by means of a wick or otherwise, the temperature will continue to fall until it approaches the *dew point.* This is nothing other than the boiling point of water corresponding to the pressure of the aqueous vapor in the air. In fact, the humidity of the air can be measured in this way, as will be more fully explained in Chapter XIII. On the other hand, when radiation falls upon a thermometer, it registers a higher temperature than that of the surroundings, as will be explained in Chapter XV. A perfectly correct thermometer therefore does not always register the temperature of its surroundings, and special precautions have sometimes to be taken to insure that it shall do so.

Three and a half decades after Faraday's work, the chemist Wüllner made numerous measurements of the boiling points of solutions. He found in general that the boiling point was raised in proportion to the concentration of the dissolved substance. But there were many exceptions, many substances behaving quite erratically and incomprehensibly. The presence of a dissolved substance also *lowers* the freezing point, and by a considerably greater amount than it raises the boiling point. Here, too, the effect was generally in proportion to the concentration, but there were many exceptions.

This whole subject remained in the greatest confusion until Raoult in 1883 found that *organic* substances behaved quite regularly. For nearly all of these, the freezing point depressions were strictly proportional to the concentrations. Furthermore, he found that if we take of each of these substances a number of grams equal to its molecular weight—what is called a *gram-molecule*—and dissolve it in one hundred grams of water, then all of these solutions show the *same* freezing point depression, about 33 Fahrenheit degrees. But all gram-molecules contain the same number of molecules. Hence we may say that the freezing point depression per dissolved molecule is the same for all organic substances dissolved in water, or the depression is proportional to the molecular concentration. This is called Raoult's law, and the depression produced by a gram-molecule is called the *molecular depression*. He found the same law to hold for other solvents, but each gives a different molecular depression.

There were still some exceptions to this rule, but Raoult found that they all could be laid to two causes. The first of these is *hydration*, the chemical union of each molecule of the substance with one or more molecules of the solvent, which, by depleting the solvent, has the effect of increasing the concentration. The second is *polymerization*, the combining of the molecules of the substance with each other into groups of two or more. Since each group acts as a single unit, this has the effect of reducing the concentration. What counts, therefore, is the number of independent parts into which the dissolved substance is divided, in proportion to the number of free parts of the solvent present.

In 1888, Raoult made a similar investigation with regard to the boiling points, and, so long as he confined himself to organic substances, found a similar rule to apply. If a gramm-molecule of each of these is dissolved in one hundred grams of water, each produces the same boiling point elevation, about 9 degrees Fahrenheit. This is the *molecular elevation* for water. Other solvents give other elevations. The few exceptions were again found to be due to hydration and polymerization.

The two phenomena are really one. What a dissolved substance actually does, Raoult showed, is to lower the vapor pressure of the liquid, and he was able to sum up all his results in a single statement, as follows: If any organic substance is dissolved in any solvent, in the proportion of one molecule of the former to one hundred molecules of the latter, the vapor pressure of the liquid is lowered by one per cent. This occurs not only at the boiling point, but at every other temperature. In short, the whole steam line is lowered, as shown in Figure 28. The full line is the steam line for water; the dotted line is that for a 20 per cent solution of sugar in water, which is about in the proportion of one molecule of sugar to one hundred molecules of water. The figure is not drawn to scale, and the depression of the steam line is much exaggerated.

Now water boils at 212 degrees because at that temperature the vapor pressure within the liquid reaches the value of one atmosphere, and is then equal to the external pressure. Under these circumstances a bubble of steam can form in the liquid. When a substance is dissolved in the water, the vapor pressure of the liquid is lowered, and at 212 degrees is less than one atmosphere. Hence the temperature of the solution must be raised until its vapor pressure reaches this value. Thus in the figure, *B* is the boiling point of water at 212 degrees and one atmosphere. In order to boil the solution, its temperature must be raised until its steam line intersects the constant pressure line *PB* at *B'*, an elevation in this case of three-tenth degrees. Similarly a solid melts when its vapor pressure becomes equal to that of the liquid, in short, when the frost and steam lines

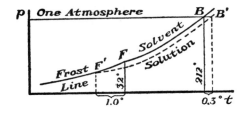

FIG. 28—STEAM LINES OF SOLVENT AND SOLUTION

intersect. This occurs at *F* for the liquid, but at *F'* for the solution, or one degree lower in this case. From the figure one can see also that the considerable lowering of the freezing point is due to the fact that the slopes of the frost and steam lines differ but little.

But why is the vapor pressure of a liquid diminished by a dissolved substance? The answer to this question has come to us by a most curious and circuitous route. Every time we drop a spoonful of sugar into a cup of coffee, a marvelous thing happens, but it is so commonplace that it excites no wonder. The sugar, of course, sinks to the bottom because it is heavier than the liquid. But it does not stay there. In a little while, even if we do not stir the coffee, we find that some of the sugar has risen to the top, for by sipping a little coffee from the surface we find that it is sweet. After a longer time, we shall find that the whole cupful is uniformly sweet. The sugar has distributed itself uniformly throughout the liquid. And it will remain so distributed no matter how long the solution stands. If we pour a spoonful of sand into the coffee, it too sinks to the bottom. But it *stays* there. If we pour in a spoonful of oil, it rises to the top, for it is lighter. And it stays there. Neither of these substances will diffuse throughout the liquid.   Why, then, does sugar do so, and how can it raise itself against the force of gravity?   Why does sugar and not sand possess this ability? It explains nothing to say that the one is soluble while the other is not, for by solubility we mean the very power we are speaking of.   We do not solve a problem by giving it a name.

Let us examine the facts a little more closely.   In the year 1819 Gay-Lussac had a dispute with some of his colleagues, who contended that if a solution were allowed to stand long enough the dissolved substance would finally settle down, and the solution become more concentrated at the bottom.   To test this assertion, Gay-Lussac filled several tubes over six feet in length with different solutions, and having sealed them hermetically to prevent evaporation, set them in an upright position and allowed them to stand for *six months*.   At the end of that time he took samples from the top and bottom of each tube, and, by the most accurate analyses of which he was capable could find no difference in the concentrations at the top and bottom of any tube.   Hence the diffusion is both uniform and lasting.   According to modern theory, however, the concentration at the bottom should be slightly greater than at the top, but it would take a much longer tube to show it.   At a depth of 300 feet, a 10 per cent solution of common salt should have a concentration one twentieth of one per cent greater than at the top.   The reason for this will appear presently.

The slow diffusion of a dissolved substance throughout a liquid can be observed by dropping a crystal of some highly colored substance into a bowl of quiescent and preferably hot water.   Permanganate of potash, obtainable at any drug store, is excellent for this purpose, because it is highly soluble and has an intense purple color.   Take one of the tiny crystals and let it slide down the side of the bowl into the water.   As soon as it touches the water it begins to dissolve and, as it slides toward the center of the bowl, it leaves behind a trail of what looks like purple smoke.   In fact, it looks for all the world like a miniature destroyer laying down a smoke screen.   This is a concentrated solution, which being heavier than water slides slowly down toward the center of the bowl.   Meanwhile the little crystal at the bottom continues furiously to belch forth fumes until it is completely dissolved.   These fumes form a little pool at the bottom of the bowl.   For a few minutes it looks as though all

were over. But if we continue to watch, we shall presently see
that the surface of the purple pool is slowly rising. It becomes
arched, its color less intense, its outlines less distinct. Soon it
becomes nearly globular in form. Hazy and cloudlike, it
slowly expands until it fills the whole liquid, and the purple
tinge is uniform throughout.

Suppose we wished to stop this expanding cloud. A sure
way would be to stretch an impenetrable barrier across the
liquid. When the cloud reached this, it would of course stop,
just as it did when it reached the sides of the vessel and the
surface of the liquid. But suppose we stretch what is called a
semipermeable membrane, that is, one which stops the dissolved
substance, but allows the liquid to pass freely. Then a curious
thing happens. The liquid seeps through the membrane into
the solution below, as though it has a strong attraction for it.
This causes the solution to expand and press up against the

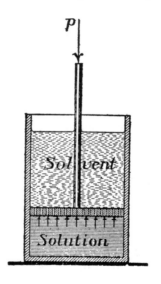

FIG. 29—OSMOTIC PISTON

membrane with a considerable force. This pressure, which
can be measured, is called the *osmotic pressure*.

Suppose we stretch the membrane across the bottom of a
piston pierced by many holes, and fit the piston in a cylinder,
as shown in Figure 29, with the solution below and the pure

solvent above.   The expanding cloud of the dissolved substance
will slowly push the piston right up to the top of the liquid.
And if we elongate the cylinder and add more liquid, the piston
will be pushed up to the top of that, and so on indefinitely.
Of course you can prevent the piston from rising by pushing
down on it.   But you will have to push surprisingly hard to
stop this fluffy cloud.   A 30 per cent solution of sugar in water
will exert an osmotic pressure of 20 atmospheres or 300 pounds
to the square inch.   You might as well try to stop the piston of
a locomotive.   And there are solutions which give osmotic
pressures as high as four or five thousand pounds.   It is no
wonder, then, that a substance in going into solution can lift itself against gravity, when it can lift in addition loads so much
greater than its own weight.

Curiously enough, the first extensive and accurate measurements of osmotic pressure were made by a botanist, Pfeffer, in
1877.   This happened because the matter has much to do with
the exchange of liquids between the living cell and its environment.   Pfeffer's work was naturally confined mostly to organic
substances.   For these he found that the osmotic pressure was
proportional to the concentration and to the temperature.

As the piston rises in Figure 29, the solution dilutes itself, and
the pressure diminishes.   When the piston has risen to the
double height, the volume of the solution is doubled and its
concentration is reduced to one half.   According to Pfeffer the
osmotic pressure is then also reduced to one half.   Hence the
pressure varies inversely as the volume of the solution.   If the
concentration, that is the volume, is kept constant, the pressure
varies directly as the temperature.   Is there not something
familiar about these two statements?   Of course there is.
They are nothing other than Boyle's and Gay-Lussac's laws for
perfect gases.   Osmotic pressure therefore obeys the same laws
that gas pressure does.

Indeed, there is much similarity between the process of dissolving and that of fusing.   Only, instead of the molecules
flying apart from excess of heat agitation, they seem to be pried
apart by the penetration of the liquid into the solid.   This is
heavy work, and demands energy.   A latent heat is therefore
required, just as for fusion, only in this case it is drawn from the
liquid, which is thereby cooled.   Thus, if sodium sulphate is
dissolved in water that is already tolerably cold, a drop in

temperature to 5 degrees Fahrenheit can be produced. The molecules of the solid, once freed from one another's grasp, fly about after the fashion of gas molecules, but are hampered in their movements by the surrounding molecules of the liquid. Nevertheless, they gradually bump their way through these, and eventually fill completely and uniformly any volume of solvent that is offered to them, just as a gas expands and fills any available volume of space. The only difference is that a substance diffuses slowly, whereas a gas expands rapidly. But the effect in the end is the same.

This analogy between a dissolved substance and a gas was first pointed out by Van't Hoff in 1887, and was pursued further by him. If we take a gram-molecule of various organic substances, he showed, and dissolve each in the same quantity of water, then each of these solutions will, at the same temperature, show the same osmotic pressure. But all of these solutions contain the same number of molecules. Here we have a perfect analogue of Avogadro's fundamental gas law, which states that equal volumes of all gases at the same temperature and pressure contain the same number of molecules. Furthermore, this analogy holds not only qualitatively but quantitatively. A given number of dissolved molecules in a given volume of solvent at a given temperature, will exert the *same* pressure as an equal number of gas molecules in an equal volume and at the same temperature. In short, if from a solution which completely fills a closed container we could remove all the molecules of the liquid, and leave those of the dissolved substance in place, the pressure exerted by the latter would remain unchanged. But in place of a solution, we would then have a gas.

We have so far compared the behavior of a dissolved substance with that of a *perfect* gas. The perfect gas laws are closely obeyed, however, only when the solutions are very *dilute*. But this need not surprise us, since gases themselves do not follow these laws very closely unless they are sufficiently attenuated. Now solutions are ordinarily much more concentrated than gases. A thirty per cent sugar solution, for example, contains in a given volume twenty times as many sugar molecules as the same volume of air contains of gas molecules at the same temperature. It is for this reason that the sugar solution exerts twenty times the pressure that the air does, or twenty atmospheres. Nevertheless, it is useful to set up the fiction and study the behavior of an "ideal solution," just as we did that of a "perfect gas," and express the behavior of actual solutions by their departures from it.

We can now see why the concentration of a solution at the bottom of a tall tube is slightly greater than at the top. We have only to suppose the liquid molecules removed and we have then simply a column of gas. The layers of gas at the bottom will be slightly compressed by the weight of those above, and will be denser, just as the atmosphere is denser near the earth. Indeed, the increase in concentration of a solution with the depth can be calculated from the same formula used for the air.

We can now at last see why the vapor pressure of a liquid is lowered by a dissolved substance. Let us pour a quantity of some solution into a thistle tube, close the mouth with a semipermeable membrane, invert the tube and dip it into a vessel containing pure solvent, say, water, as shown in Figure 30.

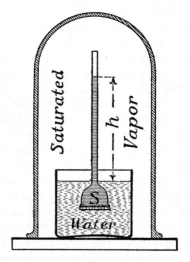

FIG. 30—OSMOTIC AND VAPOR PRESSURE

Water then penetrates the membrane and passes through into the solution. The volume of the latter is thereby increased, and it rises in the stem of the tube. This goes on until the weight of the column balances the effort of the water to get in. The height of the column then measures the osmotic pressure.

Suppose, now, we cover the whole apparatus with a bell glass jar hermetically sealed as shown. The space within soon becomes filled with saturated water vapor. This vapor is in equilibrium both with the water in the vessel, and with the solution at the top of the column, for evaporation has ceased at both places. Now the pressure of the vapor at the top of the column is less than it is at the level of the water by the weight of a column of vapor equal to the height of the solution column, h in the diagram. Hence the vapor pressure of the solution must be reduced by a like amount, for otherwise evaporation would occur at the top of the column, and it does not. Knowing h, we can calculate the amount of this reduction. Hence, from the osmotic pressure of a solution, its vapor pressure at any given temperature, and consequently its boiling and freezing points, can be calculated. It follows that all solutions having the same osmotic pressure have also the same boiling and freezing points—*isotonic* solutions they are called. Such solutions when separated by a permeable wall do not interpenetrate. Thus the surgeon uses a solution that is isotonic with the blood serum to wash a wound, for it does not penetrate the tissues and dilute the blood, nor draw the blood out and increase the bleeding.

But all these beautiful theories of Raoult and Van't Hoff applied only to *organic* substances. All the common *inorganic* substances, which comprise by far the greater part of our material world, refused to obey them. They gave osmotic pressures, boiling and freezing point changes, that were too large—that corresponded to more molecules than were actually present. Sometimes they gave values that were twice too large, as though each molecule had been split in two. This was a very serious difficulty which threatened the whole theory. The manner in which it was finally overcome is also curious and roundabout.

As early as 1878, a young student, Svante Arrhenius, at the University of Upsala in Sweden, had conceived a new theory as to the way in which a current of electricity is conducted through a solution. Shortly afterwards he offered a dissertation on the subject as candidate for the doctor's degree. This dissertation involved both physics and chemistry, but the science of physical chemistry was yet unborn—at least no one then knew that this paper was its birth certificate. There was a Department of Physics and a Department of Chemistry, but neither wished to father the hybrid child. The chemists particularly objected to the proposal of this young candidate, that when a substance like sodium chloride (common salt) was dissolved in water, the molecule NaCl, one of the tightest combinations of two elements known and requiring a great deal of energy to separate them, split in two, and formed a sodium and a chlorine *ion*, the one positively and the other negatively electrified. Now sodium and chlorine are also among the most active elements known. Sodium has a strong affinity for water, and unites with it with flame and fury, while chlorine is a pungent, poisonous yellow gas. How could the mere addition of electrical charges tame these furious elements, and make them to lie down peacefully together in a beaker of water? But the physicists saw that this theory accounted very handily for the conduction of electricity through a solution. The two sets of charged ions move, by virtue of electrostatic attractions, in opposite directions, and transport the electricity from pole to pole. Many obscure details of electrolytic conduction also became plain.

Arrhenius got his degree, but a full account of his *dissociation theory* was not published until 1888. Meanwhile he had worked with Van't Hoff and shown him how the splitting of the molecules accounted for the abnormally high osmotic pressures of inorganic solutions, for these are all conductors of electricity, whereas organic solutions are nonconductors. The dissociation, however, is not always complete. When a solution is very concentrated, only a few molecules split. As the solution is increasingly diluted, more and more of them split. But the degree of dissociation, Arrhenius showed, can always be determined by measuring the electrical conductivity of the solution, and when in this way are counted all the independent parts, molecules, groups, and ions, into which the substance is divided, it obeys the laws of Raoult and Van't Hoff perfectly.

It is perhaps needless to add that the dissociation theory of Arrhenius did much more than solve this one problem. It gave birth to the new science of physical chemistry; it was the beginning of the new electrical theory of matter; it gave a new conception of chemical affinity; and it was, so to speak, the entering wedge that finally split the atom itself into protons and electrons. The revolution it effected in chemistry was only the forerunner of the still greater revolution that followed in physics.

Avogadro's law enables the chemist to determine the molecular weight of any gas from its density, for the one is proportional to the other. Van't Hoff's extension of the law to solutions enables the chemist to determine the molecular weight of any substance that can be dissolved in any solvent. For this purpose he measures either the boiling point elevation or the freezing point depression, preferably the latter, because it is larger. The method is extremely useful, especially in the case of organic substances with their huge and complicated molecules, containing sometimes a thousand or more atoms. So if the physicist has rudely upset the chemist's ideas with his dissociation theory, he has in return provided him with a powerful weapon for knocking his molecules to pieces, and counting the number of atoms they contain.

The depression of the freezing point of water by dissolved substances is the basis of many freezing mixtures, such as that of ice and salt used in making ice cream. The salt dissolving into the ice lowers its melting point, thus causing some of it to melt. But this requires that the latent heat of fusion be supplied, and this must be drawn from the mixture itself and its surroundings. And so the melting and cooling proceed hand in hand, until with a proper proportion of the ingredients a temperature of o degree Fahrenheit can be attained. Curiously enough, the amount of heat in the mixture remains unchanged, except for what is absorbed from the surroundings. A part of it merely passes from the temperature-producing kind into the latent form, with a consequent reduction in temperature.

There are other and more powerful freezing mixtures. One part of calcium chloride to four of shaved ice will produce a temperature of − 65 degrees Fahrenheit.

When a dilute solution is cooled to its freezing point, pure ice forms at first, just as pure steam is boiled off from a solution. Thus, the ice formed on the ocean is not salt. As more and more ice separates out, the remaining solution becomes more and more concentrated, and the freezing point descends until the solution has become saturated. Then, and then only, the solution itself freezes. Salt ice can be produced in this way only at a temperature of − 8.6 degrees Fahrenheit.

XI

REAL GASES

THE reader who had a tough time in Chapter III, because of too many dimensions and other complications, will have a worse time in this one. We there merely mentioned the fact that there were complications, that real gases did not conform entirely to the patterns of perfect behavior laid down by Boyle and Gay-Lussac. We now propose to set forth what some of those complications are. We do this, not from any desire to be annoying, but because some of the most remarkable discoveries have resulted from the study of the bad behavior of gases and from the endeavor to find out just why they behaved badly. We wish to tell about these discoveries.

We pointed out in Chapter III how preposterous it was to suppose that any self-respecting gas would suffer itself to be crushed out of existence by any excess of pressure whatsoever, as according to the rule of Boyle it should. A real gas is composed of molecules of finite size, and, when these are finally pushed into contact, any further reduction of volume must encounter an enormously increased resistance, and this may set in quite suddenly, as when a fellow is pushed against a wall. However, when the first definite deviations from Boyle's law were found by Regnault in 1847, they turned out to be just the other way around. Instead of a greater resistance at high pressure than the law demanded, less was found. Instead of stiffening enormously, the gases weakened, and the change, instead of being abrupt, was gradual. The effect was most pronounced for carbon dioxide, very slight for air and nitrogen, absent for hydrogen. All this was very puzzling.

The matter was finally cleared up by a remarkable series of experiments begun by Thomas Andrews in 1863 on carbon dioxide. The accidental liquefaction of chlorine gas by Faraday in 1823 had led to the discovery that many gases and vapors could be liquefied at ordinary temperatures by pressure alone. All efforts, however, on the part of Faraday and of others to liquefy oxygen in this manner failed.   And of course no success was had with the other so-called permanent gases.   Of the liquefiable gases, carbon dioxide was one of the hardest.   It was liquefied by Thirlorier in 1834 at a pressure of sixty atmospheres.

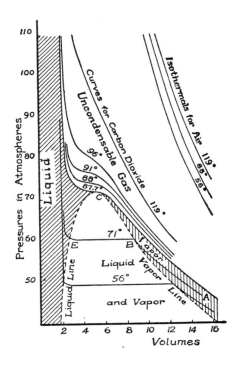

FIG. 31—ANDREWS'S ISOTHERMALS FOR CARBON DIOXIDE

Andrews compressed this gas in a narrow glass tube, at various constant temperatures up to a pressure of four hundred atmospheres, and measured the corresponding pressures and volumes.   The temperature ranged from 52 to 119 degrees Fahrenheit.   At times the contents of his tube were partly or wholly liquid, at other times wholly gaseous.   He plotted the isothermals thus found, several of which are shown in Figure 31.

One may see at a glance that the isothermals of Andrews are

very different in form from those of Boyle, as given in Figure 16, Chapter III—in some cases, indeed, having scarcely any resemblance to them at all. For comparison, three isothermals for air are given, and appear in the upper right-hand corner of the diagram. These obey Boyle's law almost exactly. How close together they are! How widely separated on the other hand are the isothermals for carbon dioxide, for the *same range of temperatures*!

The highest isothermal for carbon dioxide, that for 119 degrees Fahrenheit, resembles, indeed, those for air, but is less steep. This means that, as we proceed up the curve in the direction of increasing pressure, the volume diminishes faster than according to Boyle's law. The gas offers less resistance to compression than it should, as was found by Regnault. The next four isothermals bend first to the left, and then straighten up. This means that there is first a less, then a greater, resistance to compression than required by Boyle. Finally these isothermals rise almost vertically, showing an incompressibility almost as great as that of a liquid, notwithstanding the fact that the substance is still wholly gaseous.

The next lower isothermal at 71 degrees, however, is of an entirely different character. Starting at *A*, far down to the right, and proceeding up the curve, we find that the pressure rises as the volume diminishes, as we should expect of a gas. But at *B* a sudden change takes place. The ascent of the isothermal stops. It turns and runs horizontally to the left, along a constant pressure line. This means that no additional force is now required to reduce the volume by the very large amount represented by *BE*. At *E* conditions are just as suddenly reversed. The line turns sharply upward, and ascends almost vertically. This means that an enormous resistance is now opposed to further reduction of volume. The substance has become almost incompressible. What happened was that at *B* the gas began to liquefy. This proceeded regularly along *BE*, the quantity of gas diminishing, while the quantity of liquid increased. At *E* the liquefaction was completed. Hence a mixture of the two states was present all along *BE*. This is simply our deep cistern barometer experiment of page 70, which the reader will recall showed that, when the temperature is kept constant, the condensation of a vapor takes place at constant pressure. Indeed, the method of Andrews was very similar, for he compressed his gas by forcing a column of mercury up a narrow glass tube in which the gas was confined. His original apparatus is still preserved in the Victoria and Albert Museum in South Kensington, London.

The process is reversible, as we saw in our barometer experiment. If we start with the liquid in the condition at E, and withdraw the piston or the mercury plug, the liquid vaporizes without change in pressure. After the liquid is entirely vaporized, further enlargement of the volume is accompanied by a decrease in pressure, and we descend the curve BA.

All of the isothermals below 87.7 degrees act in this way. All have straight portions along which liquefaction or vaporization takes place. If we draw a line around these portions, like the dotted one in the figure, we outline a hump-like region, within which the substance is always partly liquid, partly vapor. The left-hand side of this hump we call the liquid line, for to the left of it the substance is wholly liquid. The right-hand side we call the vapor line, for to the right of it the substance is wholly vapor. The hump region, being broad at the bottom and narrow at the top, shows that the difference in volume between the liquid and vapor diminishes as the temperature, and consequently the pressure, rises.

None of the isothermals above 87.7 degrees enter the hump region, and consequently the gas cannot be liquefied at these higher temperatures, no matter how much pressure is applied. This was a discovery of prime importance. The isothermal at 87.7 degrees, which just touches the top of the hump, and thus divides the liquefiable from the nonliquefiable gas, Andrews called the *critical isothermal*. At this point C we have the critical temperature, the critical pressure, and the critical volume of the gas. It is the same point at which the Regnault steam line ends. This line is simply the Andrews hump region seen edgewise, from the reader's right, as will be more fully explained later.

Above the hump, the critical isothermal alone divides the liquid from the gas region. The former has been shaded on the diagram. One should note how closely the isothermals immediately above the critical temperature skirt the liquid coast line, and eventually run almost parallel with it. The gas may here be compressed to a lesser volume, that is, to a greater density, than the liquid has at a lower temperature. Yet it is still gas.

The only way to liquefy the gas at these high pressures is to lower the temperature. That is, we must turn to the left, cross the isothermals, and run toward the liquid coast line.

This explains why all the attempts to liquefy the permanent gases by pressure alone had failed. They were above their critical temperatures. In fact, a few years later, in 1877, came almost simultaneously announcements from Cailletet and Pictet of their success in liquefying oxygen, by a combination of pressure and cold. The critical temperature of this gas was found to be − 180 degrees Fahrenheit.

Since then, all the "permanent" gases have been liquefied, and even solidified. The critical temperature of nitrogen is − 231, that of hydrogen is − 390, and that of helium is − 450 degrees Fahrenheit. Since the absolute zero is at − 460 degrees, helium has only ten degrees in which it can exist as liquid or solid. With these critical temperatures go the critical pressures, which are: for oxygen 50 atmospheres, for nitrogen 33, for hydrogen 30, and for helium only 2.3. It should be noted that these pressures are not high, and any greater pressures do not help liquefaction in the least. Hence, it is not high pressure, but extreme cold that is required to liquefy the "permanent" gases. Indeed, with modern technic, it is now possible to liquefy all of these gases without compressing them at all, by simply cooling them to their boiling points, which are also their condensing points, at atmospheric pressure. These are for the gases mentioned in order: − 297, − 320, − 423, − 451.5 degrees Fahrenheit.

For these permanent gases we must imagine the hump of the Andrews diagram shrunk to a very narrow region in the lower left-hand corner. For helium it becomes a tiny spot indeed. On the other hand, for a substance like steam, whose critical temperature is 689 degrees Fahrenheit, and whose critical pressure is 195 atmospheres, the hump region rises and spreads out far beyond the limits of this diagram. It has a huge area. For substances that are ordinarily solid, like iron, the critical temperatures and pressures of their vapors are still higher. There is hence a tremendous variation in the height and area of this hump region for different substances.

Before the time of Andrews, the distinction between a gas and a vapor was rather nebulous. At first it was thought they were radically different things. A vapor was something that was given off by a liquid, or into which a liquid could be converted by heating. Such was steam. Conversely, a vapor could always be condensed into a liquid by cooling or compressing, which was impossible for a gas. But when it was found that many gases could be liquefied by pressure alone, it was thought that there was no distinction. Gases and vapors were the same. The idea did not work, however, in the case of oxygen. Andrews was able to give a clear distinction. A vapor, he said, is simply a gas that is below its critical temperature, and is therefore liquefiable by pressure alone. It is a gas in the region, lightly shaded with vertical lines in Figure 31, which lies between the critical isothermal and the right-hand side of the hump.

The weakening which Regnault observed in the permanent gases at high pressures, was just the beginning of the bend to the left in Andrews's curves. Had Regnault carried his pressures higher, up to eight hundred atmospheres, as Amagat did in 1880, he would have found that the gases stiffened again at the higher pressures. Hence the weakening was only temporary. An isothermal, as it passes near the critical point, bends toward it as though attracted by it, but straightens up after passing. As we recede from the critical point, the bend in the isothermals gradually smooths out, as though the attraction of this point diminished with the distance, and at last became impotent. Regnault found a considerable weakening in carbon dioxide, because he was not far above the critical point of that gas. He found less for air because he was farther from its critical point. He found none for hydrogen because he was so far above its critical point, that the bend had quite disappeared. It can be found, however, at a lower temperature.

What is the condition of a substance at the critical point? Just below, in the hump region, it is a mixture of liquid and vapor. Just above, it is wholly gas. What would happen if we passed directly through that point? It can be done in the following way:

Suppose we have a substance in the condition represented by the point A of Figure 32, a mixture of liquid and vapor at ordinary temperature and pressure. We enclose it in a strong vessel hermetically sealed, and raise its temperature. By this process we ascend the constant volume line AC, and if the heating is continued long enough we pass through the critical point

C. At any point within the hump, the proportions by weight of liquid to vapor are given by the distances of the point from the right and left sides of the hump respectively. We see that as we ascend *AC* these proportions do not change materially. Just below *C*, they are about the same as at the start—as much liquid as ever on hand. At the moment of passing through *C*, all of this liquid must be converted into gas—and uncondensable gas at that. What happens? A fearful explosion?

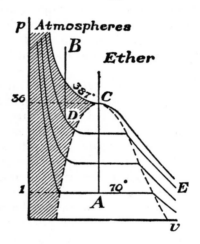

FIG. 32—HOW TO PUT A SUBSTANCE THROUGH ITS CRITICAL POINT

By no means! What is an explosion? A great and sudden rise in pressure, as in the cylinder of a gas engine, or a huge expansion in volume, as when a boiler bursts. But there is here no sudden rise in pressure. As read off on the left-hand vertical scale, the pressure rises steadily as we ascend *AC*, the rate being perfectly controlled by the rate at which we supply heat. Just above *C*, the pressure is only slightly greater than just below *C*, and has increased no more than over a stretch of similar length elsewhere along *AC*. If the vessel can withstand the pressure in this neighborhood, there is no more danger of explosion than if we had simply compressed air up to this pressure. In fact, nothing violent happens, and the change is more gradual and prolonged than one would suspect. What does happen, however, is very interesting, and cannot easily be foreseen. Let us therefore try the experiment.

For this purpose we shall not use carbon dioxide, as Andrews did, but ether, which has the advantage of being already liquid at atmospheric pressure. Its critical temperature is 387 degrees Fahrenheit, its critical pressure thirty-six atmospheres. The ether is sealed in a small glass tube, as at *A*, Figure 33, in such a way that about two-thirds of the volume is liquid, the rest vapor. The tube is then immersed in a bath of olive oil, which is heated. Since the boiling point of the oil is 570 degrees, it is well above the critical temperature of the ether at which we wish to arrive.

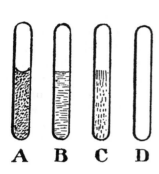

A   B   C   D

FIG. 33—THE CRITICAL TUBE

At the start of the experiment, the surface of the liquid ether is clearly seen, and is decidedly concave. This is due to *adhesion* of the molecules to the walls of the tube and to *cohesion* among themselves, as will be more fully explained in Chapter XIII. As the heating proceeds and the molecules become more agitated, they stick less readily either to the tube or to one another. Both adhesion and cohesion diminish. The curvature of the surface becomes less, until, as at *B*, the surface becomes entirely flat and barely visible. The molecules have ceased to adhere to the tube, and barely cohere to one another. As we pass through the critical point, the surface becomes hazy. It thickens into a sort of turbulent cloud belt, as at *C*. This expands and fades rapidly until the whole tube seems to be filled with a faint haze. A moment later it is clear, as at *D*. It is now completely filled with gas.

If we now lift the tube from the hot bath and allow it to cool, the reverse changes occur. A faint haze first appears throughout the tube, and seems to rain in both directions toward the middle, quickly condensing into a cloud belt, which in turn contracts until the thin flat surface is again formed at precisely the same spot at which it had previously disappeared. To see the rain ascending to re-form the cloud, and the whole series of events reversed, is like seeing a movie run backwards. As cooling further proceeds, the surface becomes more and more distinct and curved, until the original condition at *A* is restored.

During the ordinary process of boiling at constant pressure, the liquid vaporizes bit by bit, the surface of the liquid descending slowly until it reaches the bottom of the vessel. During the constant volume process that takes place in the critical tube, the surface of the liquid does not descend, but rather ascends slightly; for the density of the liquid diminishes, while that of the vapor increases, until at the critical point both have the same density, and this condition remains unchanged with further heating. Nevertheless, the approach to the gaseous state is as gradual and prolonged and even more so than in the case of ordinary boiling. It takes place throughout the mass of the liquid, and begins with the first heating. The molecules of the liquid elbow themselves farther and farther apart, while those of the vapor become more and more compacted, until at last there is no distinction between them, and the line of demarcation disappears. No latent heat is required to pass from liquid to gas in this way.

It is also possible to carry a substance directly across the coast line of the critical isothermal at a point above the hump region. This, in fact, was done by Andrews in the course of his experiments on carbon dioxide. The gas was first heated above its critical temperature; then, keeping the temperature constant, it was compressed until it was as dense as the liquid is at a lower temperature. This condition is represented by such a point as B, Figure 32. The gas was then cooled at constant volume, and thus descended the line BD into the liquid region. There is nothing spectacular about this experiment. Nothing apparently happens as the critical isothermal is crossed. There is no conspicuous change in appearance to indicate the moment of this crossing. The tube remains clear and homogeneous throughout the whole experiment. Yet at B it is filled with gas; at D it is filled with liquid. There are, however, slight changes. The liquid refracts the light, and so magnifies objects held behind it. But a conclusive proof was given by Andrews, by relieving the pressure at D while keeping the temperature constant. In this way he descended the isothermal DE. As the hump region was entered, the liquid began to boil in the usual way. Bubbles rose from its interior, the surface soon appeared at the top of the tube and slowly descended until it reached the bottom, and the space above, now much enlarged by the withdrawal of the piston, was entirely filled with vapor.

It is rather difficult to state just what the difference is between
the gas at B and the liquid at D, for both have the *same density*.
This means that the *average* distance between the molecules is
the same for both.   But at B the variations in distance are
greater.   Some of the molecules will have jostled their neighbors, for the moment, out of the way, and so have gained room
for free flights.   Offsetting this, there will be at other places
temporary overcrowdings.   But the scene is constantly shifting,
so that every molecule, at one time or another, gets a chance for
a free flight, which is the birthright of every gaseous molecule.
In the body of the liquid, the variations are less.   No molecule
ever gets away from the grasp of its neighbors.   It may wiggle a
little within that grasp, but can never indulge in a free flight.
The ultimate difference, then, between a gas and a liquid is the
presence or absence of molecular free flights.

These experiments impressed Andrews with what he called
the "continuity of state."   However greatly liquids and gases
may differ in appearances and properties under ordinary circumstances, they are, he said, ". . . only distant stages of a long
series of continuous physical changes."   They "may be made to
pass into one another by a series of gradations so gentle that the
passage shall nowhere present any interruption or breach of
continuity."   "But," he said, "a problem of far greater difficulty
yet remains to be solved, the possible continuity of the liquid
and solid states of matter."

This problem has been partially solved.   With the help of a
triple-point diagram, we can extend the Andrews isothermals
to include the solid state.   For this purpose we shall choose a
substance like paraffin which contracts on solidifying.   Figure
34 is the diagram for such a substance.   It is the same as
Figure 25 in Chapter VI, except that the "ice line" here leans
to the right, in the direction of increasing temperature, because
increase in pressure now raises the melting point.   Figure 25
was for a substance like ice that expands on solidifying, so that
increase of pressure lowered the melting point, and the ice line
there leaned to the left.

On these *pt* diagrams, an isothermal is a straight vertical line.
If the ice line leans to the right as in Figure 34, isothermals like
ABDE above the triple point, will traverse all three states.
Thus *AB* is in the vapor, *BD* in the liquid, *DE* in the solid
region.   As we compress the vapor along *AB*, liquefaction

occurs at *B* with a large reduction in volume. By further compressing the liquid along *BD*, solidification occurs at *D* with a small reduction in volume. By further compression of the solid, we simply ascend *DE*. But these volume changes of course do not show on this diagram. They do show, however, in Figure 35, which is a *pv* or Andrews diagram for the same substance, and the isothermal *ABDE* there appears with the same letters. The point *B* of Figure 34 here becomes the *line BB'*. The point *D* of 34 here becomes the *line DD'*. *BB'*

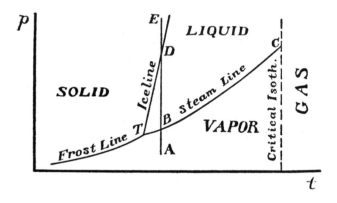

FIG. 34—TRIPLE POINT DIAGRAM FOR A SUBSTANCE THAT CONTRACTS ON SOLIDIFYING

is our familiar liquefaction line across the hump region. But we now see that the isothermal, after ascending the liquid region some distance, meets another transformation area at *D*, the solid-liquid region. Here the substance contracts as it solidifies without change in pressure. When solidification is complete, the isothermal again ascends almost vertically in the solid region.

All the isothermals between the triple and the critical points act like *ABDE*, as may be seen from the several that have been drawn in Figure 35. An isothermal above the critical point misses the liquid region entirely; but it will still hit the ice line at some stupendous pressure, provided this line continues indefinitely to lean away from the reader. We have then this curious result, that, although it is impossible to liquefy a gas that is above its critical temperature, it may be perfectly possible to solidify it!

As we go down temperature from the critical to the triple point, the liquid portions of the isothermals become shorter and shorter, until at the latter point they reduce to nothing. But the triple point, which is T in Figure 34, becomes the triple *line* TT'' in Figure 35, and a very long one at that, for it includes not only the contraction from vapor to liquid TT', but also the additional contraction from liquid to solid $T'T''$. The liquid-solid transition area springs from this segment $T'T''$, and runs

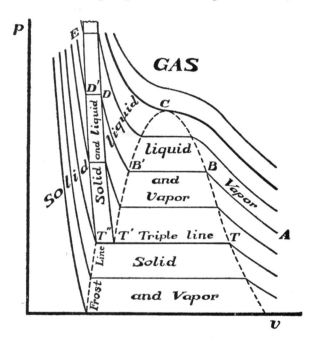

FIG. 35—THE ANDREWS ISOTHERMALS EXTENDED TO THE SOLID STATE

like a narrow ledge up the diagram. Every liquid isothermal that meets this ledge drops down it to the solid region below— or, rather, to the left. Below the triple line, the widened hump region represents a mixture of vapor and solid, and the direct transition from one to the other.

For a substance like ice that expands on freezing T'' would lie to the right of T', and the area below the triple line would be narrowed instead of widened. The solid-liquid ledge would still spring from $T'T''$ as before, but since it would lean toward instead of away from the reader, the isothermals that traverse the liquid region would not meet it. Only those below the triple line, which run up the ice region, would meet it, and would then drop off into the liquid region, which now has the lesser volume. But because the projecting ice would hide a good part of the liquid region, this case cannot be satisfactorily represented on a diagram such as Figure 35.

The complete Andrews diagram is thus seen to consist of three regions representing vapor, liquid, and solid, and three transition areas, where the changes from one to the other take place. The Regnault diagram consists of the same three regions, and three transition *lines*, where the same transformations take place. Evidently the two diagrams are only different presentations of the same thing. In fact, the one being a *pv*, the other a *pt* plot, they are related in the same way as the Boyle and Gay-Lussac diagrams of Chapter III, and may be combined into a single three-dimensional model in the same way.

To do this, we must imagine a temperature axis starting from the lower left-hand corner of Figure 35, and receding from the reader, and each isothermal set back a distance proportional to its absolute temperature. We then see that the Regnault diagram is simply this model looked at from the reader's right. The Andrews transition areas are then seen edgewise and become the transition lines of Regnault. The isothermals where they cross these areas are seen end on and become points; the isothermals themselves become lines straight up and down. The Andrews diagram is a front view of this model.

But let us relieve our imaginations from further strain by actually constructing the model. We shall not, however, use carbon dioxide, nor the imaginary substance of Figure 35, but that substance which is most abundant and useful in this world and about which we possess the most extensive and accurate information, namely—water. This will enable us to plot the isothermals to scale. Then having set them up in the manner described, we give the model one-eighth of a turn to the left, so that the projecting ice region will not hide so much, and the result appears as in Figure 36.

This model shows in a single view all the known forms of

water, ice, and steam, including Bridgman's five forms of ice.
On account of the enormous ranges of pressures and volumes
involved, it has again been necessary to use scales that greatly
compress the higher values.   The cube roots of the temperatures, the fourth roots of the pressures, and the tenth roots of the

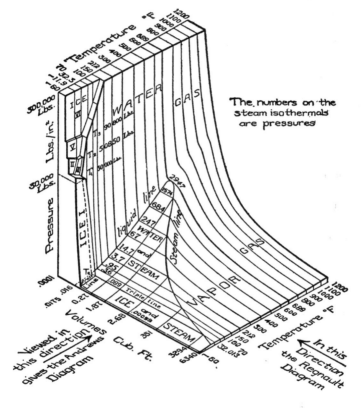

The numbers on the
steam isothermals
are pressures

FIG. 36—MODEL FOR WATER, ICE, AND STEAM

volumes have been plotted.   These scales do not destroy the
essential relations, but rather make it possible to bring them all
into a single view.   Compression of the volumes is particularly
needed.   The volume of ice at 32 degrees Fahrenheit is 0.0175
cubic feet; that of steam at the same temperature is 3294 cubic
feet, or nearly two hundred thousand times greater.   If the

former volume were represented by only one-tenth of an inch, the latter would require on the same uniform scale nearly one-third of a mile.   That would be the length of the triple line. It would be a little difficult to fold that up into a book.   On the scale chosen instead, the two volumes are represented by lengths that are in the ratio of about one to four.   Even at that, the steam line runs off the model a little below 32 degrees, so that the enormous expansion that occurs when ice passes directly into vapor at low temperatures cannot be shown.   On the other hand, the changes in volume between the different kinds of ice are so small that it has been necessary to magnify them ten times.

The model has been cut off at −60 degrees, for the scales stretch out the low regions as much as they compress the high ones, and the absolute zero lies far forward to the left.   The ice-steam area continues down to the absolute zero forming an exceedingly thin wedge.   So far as known, nothing of interest occurs in this region anyway.   The model has also been cut off at 0.0001 cubic foot.   To continue it to zero volume would only inconveniently thicken the left back wall.   The pressure or vertical axis alone has been brought down to zero.

In the upper left-hand corner of the model are Bridgman's four new forms of ice.   These are all less in volume than water, and hence become sunk in areas, separated from the water region by the ledge that runs upward and to the right.   Ice I, or ordinary ice, is represented by the pillar at the bottom.   It is really a sheet that extends forward and leftward to the absolute zero.   This ice being greater in volume than water, the ledge that separates it from the latter projects outward from the water surface.   The ordinary ice line is hidden behind the pillar, and has been shown dotted.   The ordinary triple point T_0 lies at the bottom of the pillar, T_1 at the top. T_2 and T_3 lie farther up on the ledge.   The pressures corresponding to these triple points are marked on the diagram.   They are the same points that appear in Figures 26 and 27.

It is interesting to follow the −60 degrees isothermal, which forms the left front boundary of the model.   This isothermal, after traversing the ice-steam surface, ascends the pillar of ice I, then crosses in succession ices II, V, and VI, thus crossing four kinds of ice.   At each junction there is a contraction in volume.

This model, which portrays the behavior of a real gas, should be compared with the one on page 36, which sets the standard of perfect behavior.   This one is very different in appearance and much more complicated.   Yet the complications fade out gradually as the temperature rises, and above 1200 degrees Fahrenheit the isothermals are quite smooth, and approach in form those of Boyle.

XII

BOILING BY BUMPING

THE model described in the previous chapter shows the actual behavior of a real substance in its various states. It is, however, purely empirical, that is, it is simply a résumé, an exhibition of what we know about the substance. An immense amount of experimental data was required for its construction, data that scientists have been gathering ever since the days of Regnault in 1847. We possess enough such information for a fairly complete model only for the substance water.

The model of Chapter III for a perfect gas is a purely theoretical construction. The whole thing is developed from the single equation $pv = RT$, and if all gases were perfect, we would only have to determine by experiment the value of the single constant R for each gas, and then the complete behavior of every gas under all conditions and circumstances could be calculated. That would save an immense amount of experimental work and uncertainty. Despite the great mass of data that has accumulated, it often happens, when some new problem or new industrial development comes up, that the data required is lacking, or has not been determined with sufficient accuracy, or, if it has, is so buried amid mountains of other data that it cannot be found, and new measurements have to be made. Hence the physicist would like to develop a theoretical model that approximates somewhat more closely to the actual behavior shown in Figure 36, and that does not ignore the changes of state, as the perfect gas model of Figure 18 does. Of course he cannot expect, with these added complications, to get out of it with so simple an equation as that for a perfect gas, nor to be able to make everything depend upon a single measurement for each gas. But if he could obtain a good approximation to any considerable part of the model of Chapter XI, that would take account at least of liquefaction, and would require only a few measured constants, he would feel that a great forward step had been made.

One trouble with the model of Chapter XI is its many discontinuities, that is, sharp corners. Every isothermal that passes through the hump region consists of at least three parts, one in the vapor, one in the liquid, and one in the transition region. Each part follows a separate law, hence requires a separate equation. But if these sharp corners could be gotten rid of, so that each part passed more or less gradually into the next, then the whole of such an isothermal could be represented by a single equation. The first step, then, toward our goal is to get rid, if possible, of these sharp corners, to make each isothermal a process of *continuous* change from end to end.

We believe, in fact, that this would more nearly represent what actually occurs, for we believe that nature does nothing suddenly. We have seen that amorphous substances do not have the sharp melting points that crystalline ones have, but rather places of more rapid change. It may be that the melting points of crystalline substances are not as sharp as they appear to the unaided eye, that if we could look at them with a magnifier, as it were, we would find them rounded, to be also places of more rapid but not of instantaneous change. Let us examine, then, a little more closely just what happens when a substance changes state.

Ordinarily when water is cooled, the fall in temperature stops when the freezing point is reached. Ice crystals then begin to form and the temperature remains constant so long as the freezing continues. But if pure water in a perfectly clean vessel is cooled very quietly, it is possible to reduce the temperature much below the freezing point without solidification occurring. This is called *undercooling*. Despretz succeeded in this way in cooling water to -4 degrees Fahrenheit without freezing it. The undercooled water continued to expand and to increase in viscosity as the temperature fell, just as it did in approaching the freezing point. If during the process, however, there is the slightest jar, or a bit of ice is dropped into the liquid, a considerable amount of water freezes at once, and the temperature immediately rises to the ordinary freezing point. This rise is due to the latent heat released by the sudden solidification, which warms the liquid.

An undercooled liquid is thus in a state of unstable

equilibrium, which grows more acute as the temperature falls. With most liquids, a point is at last reached where spontaneous crystallization occurs, whether there is a disturbance or not. The crystals start at innumerable points distributed throughout the liquid and grow rapidly out from these as centers until the whole mass is solid. This process has been followed with the microscope. The number of points and the rate of growth are characteristic of the substance. If solid impurities are present, the crystals form around these as nuclei, and the process begins much sooner.

It appears, then, that the freezing of a liquid at its freezing point, though the *usual*, is not the *normal* process. It is due solely to impurities and disturbances. The normal course for the pure and undisturbed liquid is to cool further to the temperature of spontaneous crystallization, and then solidify rapidly *en masse*. And the latent heat of fusion is not released until this crystallization occurs, no matter how much the liquid has been undercooled. This normal behavior is unusual only because impurities and disturbances are usual.

It is further apparent that, because of the impulse and time required for the growth of a crystal, all freezing, even that of the usual sort, must be accompanied by some undercooling, although it may be very small and occur at innumerable points. A crystal simply will not form unless there is a nucleus, or a definitely indicated point for starting, and an impulse provided by the tension of an undercooling. In fact, when measuring melting points, or when fixing the zero point of a thermometer by a mixture of melting ice and water, it is important to stir the mixtures constantly, in order to avoid appreciable undercooling at any point.

Chemists are familiar with a similar phenomenon when a substance crystallizes out of solution. A pure saturated solution may be quietly cooled, and thereby become supersaturated. But if a crystal of the substance is dropped into the liquid, a considerable amount crystallizes out at once, and the temperature rises to that at which the solution is again saturated.

While it is difficult to undercool water and substances like it to any great extent, there are others for which it is difficult and even impossible to avoid undercooling. With all substances the tendency to crystallize increases as the temperature is lowered under the ordinary freezing point. But with these other substances the tendency reaches a maximum value at some point, and thereafter diminishes as the temperature is further reduced.   If this perilous stretch can be safely run through, by rapid and quiet cooling, the temperature can then be lowered indefinitely without further danger of crystallization occurring.   Nevertheless, the substance does not remain liquid. It solidifies with its molecules, so to speak, in place, precisely as they were in the liquid state.   This process is analogous to the liquefaction of a gas with its molecules in place, as was described on page 104.   And as in the latter case there was no conspicuous change in appearance as the gas became liquid, so in this there is no conspicuous change as the liquid becomes solid.   There is no sudden change in volume, no release of a latent heat.   The properties of the substance change continuously, though more rapidly than at other temperatures.   The substance simply becomes more and more viscous until it becomes hard.   The solid is isotropic and clear like the liquid, only it refracts the light more strongly.   It is hard, brittle, and when broken the fracture is smooth and without grain or facets. The substance is a glass.

That glass is simply an undercooled liquid, that has belatedly solidified, is shown also by the fact that it sometimes does crystallize and solidify prematurely, much to the chagrin of the glass maker.   This crystallization can, indeed, be induced by the addition of certain substances called mineralizing agents. They convert the glass into an opaque crystalline mineral.

In the same way that a liquid can be quietly cooled below its usual freezing point, it can also be heated above its usual boiling point.   Thus if pure water, from which the dissolved air has been driven off by previous boiling, is quietly heated in a clean flask, especially if the heat is applied from all sides by immersing the flask in oil bath, the temperature can be raised much above the boiling point without boiling occurring.   Dufour, by heating small drops of water immersed in a mixture of oils of the same density, was able to raise the temperature to 317 degrees Fahrenheit without their boiling.   If, however, the superheated liquid is disturbed, or a foreign object dropped into it, a large bubble of steam forms all at once and the temperature drops to the boiling point.   Bubbles of steam will then continue to rise from the foreign object introduced, and if this has any sharp edges or corners, the bubbles form more readily at these places, and are then smaller and more numerous. If no foreign object is introduced, the temperature will continue to rise until suddenly an enormous bubble is formed and ejected with explosive violence. Then the liquid quiets down, the temperature again rises until another big bubble is formed and violently ejected—and so on. This process is called *boiling by bumping*. It got its name from the fact that it sometimes occurs in steam boilers, when, from the clogging of the tubes or other causes, proper circulation of the water is prevented. The intermittent boiling produces sudden rises of pressure, and a bumping noise. It is not good for the engine and may be dangerous. It signifies that the boiler should be cleaned at once.

This boiling by bumping, like the undercooling, though not the usual process, is nevertheless the normal one. The ordinary quiet boiling is due to disturbances and foreign bodies. Fortunately our pots and kettles are sufficiently rough to provide innumerable points where bubbles may easily form, and so the boiling is quiet. But the phenomenon is not so rare in the case of viscous liquids, which do not circulate freely. Watch a pot of thick porridge on the stove. Note how the middle slowly rises in a hump, breaks, and emits a big puff of steam from a crater-like opening, then subsides only to repeat the process over and over. The housewife knows that it is high time to apply artificial circulation in the form of vigorous stirring, else the porridge will burn. By so doing she keeps the bubbles small. They are then more frequent, and the rise in temperature between bumps is reduced.

In the end, all boiling is by bumping. Every individual bubble, however small, involves an overheating of the fluid in its neighborhood, a rapid conversion of a finite quantity of it into steam, and a consequent drop in temperature to the normal boiling point. Every bubble requires something on which to form. None forms in the middle of the liquid, unless there are solid particles floating about. Otherwise they all form on the bottom or sides of the vessel, especially where there are rough places or projecting points. If there is a sufficiency of these, the boiling is quiet and steady.

Because some slight bumping and overheating are always present even during the quietest boiling, when measuring the boiling points of liquids or determining the steam point of a thermometer, we place the thermometer, not in the liquid, but in the vapor just above it.   This has the same temperature, but is not subject to the fluctuations caused by bumping.   In the case of solutions, however, we cannot do this, because the thermometer then registers the boiling point of the solvent, not of the solution, as already explained.   The instrument must be placed in the liquid.   To insure, then, as even and quiet boiling as possible, a quantity of broken glass or crockery or sand is placed in the solution.

Similar phenomena occur during condensation.   The pure vapor can be compressed beyond the saturation pressure, or the temperature reduced below the dew point, without condensation occurring.   But if smoke, dust particles, liquid particles, or ions (electrified atoms), are introduced, a cloud immediately forms. The vapor condenses on the foreign particles as nuclei, and they seem to be necessary to the formation of a cloud.

These facts have led some people to believe that by discharging smoke, dust, or electricity into the air, rain could be produced.   But this can only succeed if the atmosphere is already supersaturated, in which case it will presently rain anyway, and the rain maker is unnecessary.   The method can sometimes be used, however, to precipitate a fog.

Another favorite method has been to shoot bombs up into the clouds, with the idea that the explosions would somehow coagulate the droplets, instead of scattering them as might more reasonably be expected.   The idea seems to be based on reports of rains that have followed battles, or on the observation that rain frequently falls just after a heavy clap of thunder.   With regard to the latter, it must be remarked that, since the sound reaches us very quickly, while the rain takes some time to fall, it is more likely that the rain started first and provided the semiconducting path that enabled the big flash to occur.   The cart has been put before the horse.   With regard to more occult methods of producing rain, nothing need be said.

In 1871 James Thomson, the brother of Lord Kelvin, suggested that these phenomena of supersaturation, etc., would be more closely represented by isothermals of the form shown in Figure 37, than by those of Andrews.   Starting with the vapor at A, and compressing isothermally along AB, condensation does not begin at B, but the pressure continues to rise until D is reached.   Then a considerable amount of vapor condenses all at once, while the pressure falls and then rises again finally to the value corresponding to the boiling point. Or starting with the liquid at *H*, and heating along the constant pressure line *HKL*, boiling does not begin at *K*, but the temperature continues to rise, for the line crosses the higher isothermals until we come to the center of the depression at *L*. Then we recross the same isothermals in the reverse order. The temperature is now falling, and vaporization occurs.

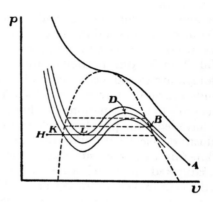

FIG. 37—JAMES THOMSON'S ISOTHERMALS

These isothermals of Thomson are continuous curves. Hence they are such as can be represented throughout their entire lengths by a single mathematical equation. If an equation could be found to fit them and also conform to the requirements of gases, the aim of the theoretical physicist mentioned at the beginning of this chapter would be partially realized. Such an equation, in fact, was found by the Dutch physicist, Van der Waals, in 1879. He reflected that the theory of perfect gases is based on the assumptions that the molecules have no size, and that there are no attractions between them, both of which assumptions are false. The molecules certainly have size, and determine a minimum volume for the gas. Since the molecules cohere strongly when in the liquid and solid states, there must also be attractions between them; and while these forces may be very feeble when the molecules are far apart, they doubtless become of importance when the gas is concentrated and approaches its liquefaction point. Their effect would be to diminish the pressure which the gas exerts by virtue of the impacts of its molecules on the walls of the containing vessel, for a molecule cannot strike so hard if held back by its neighbors. Van der Waals therefore introduced into the perfect gas equation $pv = RT$, two new terms to correct for its two false assumptions. These involved two new constants, so that his equation contains three constants, to be determined by experiment for each gas, instead of the one constant R of the old equation.[1]

Van der Waals's equation gives below the critical point isothermals greatly resembling those of William Thomson, and above this point isothermals closely resembling those of Andrews. If the temperature is further raised, the isothermals gradually approach the form of those of Boyle. Thus the behavior of a gas, both far above and considerably below its critical point, is very faithfully represented. But quantitatively the agreement with experiment is not as good as could be desired. Other and more complicated equations have been proposed, and some of them are more exact than that of Van der Waals; but they involve additional constants and arbitrary assumptions that have no direct physical meanings. They are empirical formulations that have simply been fitted to the facts, and so represent no theoretical advance.

A complete equation of this sort should of course include the solid state. But here the complications, with the many crystalline and amorphous forms, are much too great. Besides, we have not yet enough knowledge. A great deal of research is now being done on the solid state, aided by X-ray analysis and other powerful new weapons. But so far the results have only been to uncover endless new details and complications, which give rise to innumerable new theoretical problems. And it

[1] Van der Waals's equation is $(p+a/v^2)(v-b) = RT$; a and b are the two new constants to be determined by experiment; b is simply the minimum volume of the substance, the space occupied by it when the pressure is infinite. Then $(v-b)$ is the additional space occupied by the molecular motions. This subtracting out of b is equivalent to moving the pressure axis of the diagram a distance b to the left. The strip of space thus set off is never invaded by the isothermals produced by the equation. The term a/v^2 is the additional restraint put upon the motions of the molecules by their mutual attractions. It is equivalent to a pressure which is added to the external pressure. Since each molecule attracts every other, Van der Waals assumed that this additional restraint would be proportional to the square of the density, that is, inversely proportional to the square of the volume of the gas. When the volume becomes very large, the value of this term becomes very small, so that the isothermals of Van der Waals then approximate to those of Boyle.

appears that there is still an immense amount of fine detail that lies beyond even the reach of the X-rays, to which we have only just begun to penetrate. We cannot, of course, begin to form a comprehensive theory until a good part of the facts concerned are at hand.

I sometimes wish that those philosophers who preach so powerfully about unity, and about its desirability, as though we had only to wish for it in order to have it, would try their hands at unifying the solid, liquid, and gaseous states of matter. Unity is easy to achieve when one's eyesight is poor, and all things look alike anyway. But when one's vision is sharpened by high-powered microscopes, and by instruments and methods that probe many times more deeply, he discovers such endless differences and variations, that to bring unity into only a minute fraction of them is the work of a lifetime. The scientist who, more than any one else in the world, is laboring to bring unity out of chaos, talks very little about it. He hasn't the time.

XIII
MOISTURE

WHAT do we mean when we say that water is wet? We mean in general that it will adhere to the fingers or to other solid bodies that are dipped in it, forming a thin film over their surfaces. Other liquids are wet in the same way. On the other hand we say that mercury is dry, because it does not stick to the fingers, to glass, to cloth, or to many other objects. We usually think also that a liquid that is wet is always wet, and one that is dry is always dry. But this is not so. Water will not stick to a greased surface, but gathers itself into little globules, as though it were repelled by the grease, and were trying to get as far away from it as possible. Water runs off a duck's back. On the other hand, mercury sticks very strongly to clean metallic surfaces, and forms a film over them just as water does when it wets an object. Nevertheless, we do not call this wetting, but amalgamation—perhaps because we cannot wipe the mercury off with a rag, as we can water. But this is only because mercury does not wet the rag as water does. Obviously two substances are always concerned in the matter of wetting, the one that wets and the one that is wetted; and no liquid is universally wet, nor is any universally dry.

Wetting is due to the attraction or adhesion between the molecules of one body and those of another. This attraction is always present; nevertheless wetting does not always occur. In order for it to occur, two things are necessary. First, the two bodies must be brought into very close contact over considerable areas, because the molecular attractions are effective only at very short distances. The molecules of different solids, for example, attract one another. But the surfaces of solids, even when they appear quite smooth, are, molecularly speaking, rough. Two such bodies come into intimate contact only at isolated points. The *adhesion* produced by these few points is too feeble to be effective. But when two pieces of glass are ground optically flat, and the surfaces are perfectly clean, then when they are pressed and wrung together, they adhere with a very appreciable force. A liquid, on the other hand, flows into every irregularity of a solid, and so comes into intimate contact over large areas. Hence it can adhere strongly. If the wetted surfaces of two pieces of glass are brought together, even if their surfaces are far from being optically flat, they will adhere strongly. They will slide easily on one another, but to pull them directly apart requires considerable force.

The second condition that must be fulfilled if a liquid is to wet a solid is that the attraction between its molecules and those of the solid must exceed the attractions between its own molecules. In short, the *adhesion* must exceed the *cohesion*. For the molecules to stick to a foreign object, they must be drawn away from their fellow molecules—the foreign attachments must exceed the home ties. If the reverse is true, the liquid draws away from the foreign object, not because there is a repulsion— for there is not—but because the home ties are stronger, as illustrated by the way water on a greased surface draws itself up into globules.

A further illustration is afforded by the behavior of liquids in tubes of small bore, especially in *capillary* or hairlike tubes. If such a tube is dipped in a liquid which wets it, the liquid is drawn a short distance up the tube, not only against gravity, but against the restraining attraction of the rest of the liquid, and the surface is concave, as shown at A in Figure 38. If, on the other hand, the tube is dipped into a liquid which does not wet it, like glass in mercury, the liquid is depressed in the tube, and the surface is convex, as at B in the figure. There is still attraction between the mercury and the glass, but the attraction of the rest of the mercury in the vessel is the stronger, and is able to pull some of the mercury down out of the tube. A rag is an excellent wiper for liquids that wet it because its many interstices form so many capillary tubes into which the liquid is drawn. But it will not wipe mercury because the latter is not drawn into these tubes.

We also distinguish degrees of wetness. We say that a body is saturated, is sopping or wringing wet, when it is as wet as it can be. When less wet we call it moist or damp. But in every case we mean the presence of more or less liquid. Hence wetness is something that pertains to liquids. In the following we shall further confine ourselves to the wetness that is produced by water.

Nevertheless there is a general notion that ice and steam are also wet—in short, that water in all its forms is wet. The idea arises perhaps from the fact that all three forms wet the fingers. But ice does so because of a film of water that is produced by its melting, and this is caused either by the higher surrounding temperature or by the warmth of the fingers themselves. Ice below the freezing point is perfectly dry. On the other hand, steam condenses to water on the fingers because the latter are cooler. The steam before it condenses is perfectly dry, for if we put a dry object in it, which is at the same or a higher temperature, the object will remain dry. In both cases the wetting was done by water. Of the three forms, then, water alone is wet, and does all the wetting.

Steam in fact is a gas, as dry and invisible as air. The white clouds that we see issuing from a steam pipe are not steam, but are composed of fine particles of water into which the cooled steam has condensed.

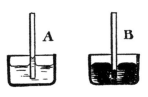

FIG. 38—CAPILLARY ELEVATION AND DEPRESSION

If we boil water in a glass flask, as in Figure 39, the interior is perfectly clear, and nothing is seen of the steam until this cloud is formed. If the steam issues rapidly from a fine nozzle, a transparent space may often be seen between the nozzle and the cloud, where the steam has not yet condensed. Afterwards the cloud disappears, which means that the water reëvaporates, and becomes again invisible steam, but at a lower temperature and pressure. All fogs, clouds, and mists are composed of fine particles of water or of ice. All vapors are dry, being merely gases below their critical temperatures. There are no "clouds of steam."

Yet the engineer often speaks of *wet steam*. Now how can steam ever be wet, if all vapors are dry, and all wetting is done by water sticking to the object wetted? Do we mean that molecules of water attach themselves to a molecule of steam and wet it? That would be absurd, because there is no difference between a molecule of water and one of steam. When the molecules are flying about freely and separately, they are steam. When a number are clumped together, they become a drop of water, and there is no distinction between the molecule that is wetted and those that are doing the wetting. Nevertheless, something of the sort we do mean. Steam is wet when a certain amount of water in the form of a fine mist is intimately mixed with it. The degree of wetness is measured by the percentage of the mixture by weight that is water.

This "mixture" must be distinguished from that represented by a point on the Regnault steam line, or within the hump region of the Andrews diagram. In this latter mixture, all of the water is at the bottom of the vessel, all of the steam is at the top. The two are *separated* by the liquid surface. Steam in this condition, in contact with its liquid, but not mixed with it, is dry and saturated— saturated, not because it contains any water, but because it is just ready to condense.

We speak also at times of a gas being moist or humid, of the air being damp. Do we mean that a certain amount of water is intimately mixed with it, as in the case of wet steam? Do we mean that when such air comes in contact with an object, it will moisten it? By no means! Such air

FIG. 39—STEAM IS INVISIBLE

will dry the object. For instance, the washerwoman tells us that the air is very damp to-day, because it takes a long time for the clothes hung out on the line to dry. Nevertheless they *do* dry. Damp air, then, has no wetting power, but only a *diminished drying power*. And this is due, not to the presence of water, but of water vapor, which impedes evaporation. The air is dry, the vapor is dry, hence the whole mixture is dry, and has drying power.

Nor is the so-called dampness measured even by the proportion of the mixture that is water *vapor*. A given mixture, when simply heated, becomes drier, in the sense that its drying power is increased, although the proportion of the ingredients is not changed thereby. Thus the laundry man, when it is too damp outside, hangs his clothes up in the hot room, where he uses the *same* air to dry them, but heats it first. Now this heating can in no wise destroy any of the water vapor present, nor convert it into anything else. Furthermore, both the air and the vapor were already perfectly dry, hence cannot be made any drier by heating. Yet the drying power of the mixture is undoubtedly increased. What is the explanation?

In the first place, the air has nothing to do with the drying. It cannot soak up water like a towel, because it cannot be wetted. Hence it can do no drying. But since evaporation increases with the temperature, hot air may promote the process by heating the objects to be dried. It may thus act as a heater, but not as a drier.

Evaporation, as we have seen, results from the livelier molecules shooting out of a liquid. But some of them shoot back again, and the rate at which they do so depends upon how many are on hand. If none are on hand, as when a liquid is first introduced into a vacuum, the rate of evaporation is then a maximum, and is determined only by the temperature. But as evaporation proceeds—the liquid being in a closed container kept at constant temperature—the density of the vapor increases, the number of molecules shooting back increases, and the net rate of evaporation *decreases*, until, when the vapor reaches the saturation density corresponding to the temperature, the net evaporation ceases. Hence the "dampness" of an environment, which is the inverse of its drying power, as measured by the rate of evaporation, increases as the vapor density increases, and reaches a maximum when the vapor is saturated. At any previous stage, the degree of dampness is measured by the quantity of vapor present in proportion to that required for saturation. This ratio is called the *relative humidity*, or simply the humidity. It is zero when no water vapor is present, 100 per cent when the vapor is saturated, 50 per cent when the vapor density is half that required for saturation at the given temperature, and so on.

In the calculations, we have paid no attention to the amount of air present, nor to the amounts of any other gases or vapors that might have been present. The only quantities used in reckoning the humidity were the amount of *water* vapor present, and the amount required for saturation at the given temperature. The air, other gases, or even other vapors that might be present, take no part in the drying process.

According to Boyle's law, the density of a gas is proportional to its pressure. This is also very nearly true for water vapor under the conditions we are considering. Now since the saturation pressure increases with the temperature, the values always being given by the Regnault steam line, it follows that larger and larger quantities of water vapor will be required for saturation, that is, for 100 per cent humidity, as the temperature rises. And the same will be true for every other degree of humidity, except 0 per cent. The same amount of vapor present will therefore produce very different degrees of humidity according to the temperature. For instance, suppose that at a temperature of 66 degrees Fahrenheit the water vapor present is saturated. The humidity is then 100 per cent. In every cubic foot there is 0.001 pound of the vapor. Let the temperature now be raised to 102 degrees. The humidity drops to $33\frac{1}{3}$ per cent, for at this temperature 0.003 pounds of vapor per cubic foot are required for saturation, and only 0.001 pound is present. Heating therefore reduces the humidity, even though the amount of water vapor present remains the same.

When the density of a vapor is below that of saturation, the vapor is superheated. The vapor in the air, then, that causes all our so-called dampness is superheated steam. It becomes at most saturated steam when the humidity reaches 100 per cent. But the reader will know full well by now that these terms are not necessarily associated with high temperatures. Steam may be saturated and even superheated at temperatures far below the freezing point.

Suppose that into this environment at 102 degrees and $33\frac{1}{3}$ per cent humidity we introduce a wet thermometer. The temperature of the latter will fall because of the evaporation of the water. But it cannot fall below 66 degrees, for at that temperature, as we saw above, the vapor is saturated, and evaporation ceases. If we introduce an object that is colder than 66 degrees, dew will condense upon it. Sixty-six degrees is therefore the *dew point* for these conditions. Having determined it in this way, we have only to look up in the steam tables the saturation densities corresponding to the dew point, and to the air temperature, respectively, and their ratio will be the relative humidity, as given above.

At low temperatures, an extremely small amount of water

vapor suffices for saturation. At 0 degree Fahrenheit the saturation density is only 0.0003 pounds per cubic foot. A small room, ten by twelve, and eight and one-half feet high, or of about one thousand cubic feet capacity, would contain 1293 ounces of air, but only 1.1 ounces of water vapor. Yet the humidity would be 100 per cent. If this room were heated to 70 degrees, the humidity would drop to 6 per cent, which is very dry. To bring the vapor again to saturation at the increased temperature would require the addition and vaporization of a whole pound—about a pint—of water. That is why the air in our heated houses is always so dry in winter, even if it is quite damp outside. Conversely, when air is artificially cooled, the humidity increases, and unless some of the water vapor is removed, the air feels uncomfortably clammy.

People sometimes find it difficult to understand how there can be "moisture" in the air, when the temperature is below freezing. They think that it should all freeze up. And so it would if this "moisture" were in the form of water globules. But it is not. It is in the form of water *vapor*, which is a gas. A liquid can form only when the molecules are brought together. In a gas they are far apart. There is hence no difficulty in cooling water vapor far below the freezing point, and still having it a gas, so long as we keep the molecules far apart. That is, we must keep the density low; we must get down under the Regnault frost line of Figure 25. This means that very little water vapor can be present at these low temperatures, for as soon as the density rises to a value on the frost line, the humidity becomes 100 per cent and the vapor condenses, not to water, however, but to frost or snow.

In all of this we have made no mention of any part played by the air. It has been entirely absent from our considerations, and might just as well have been physically absent also. The air does, however, play the part of establishing and maintaining the temperature. When a liquid evaporates, it is cooled. Conversely the vapor is warmed. The air, being of intermediate temperature, cools the vapor and warms the liquid. It thus restores to the liquid the heat it loses, and keeps up the rate of evaporation. Hence the air acts as a temperature equalizer. Other than that, it can only hinder evaporation by its molecules getting in the way of those shooting out of the liquid, deflecting some of them back, and so preventing the vapor from diffusing away from the liquid as fast as it otherwise would.   But this effect is very slight, as we shall now show.

John Dalton, the discoverer of the atomic weights, deduced as a consequence of the theory of perfect gases, that when several such gases are introduced into a closed container, each expands and fills the whole volume uniformly; and each exerts on the walls of the container a pressure which is exactly the same as it would have exerted had the others been absent. The only way the different gases affect one another is that their association constrains all to assume the same temperature. The pressure of each gas is then proportional to its density, that is, to the amount of it by weight that is present, and the total pressure on the container is the sum of the pressures exerted by the separate gases.   This is called the law of *partial pressures*. One consequence is that a small sample of the mixture taken anywhere always shows the same composition.

Now these deductions of course apply only to perfect gases, whose molecules are infinitesimal or of practically no size. Such a gas may fill a volume by its pressure, and yet occupy none of its space.   Hence when another gas is introduced, it finds the space still wholly empty.   It is not surprising, then, that several perfect gases should be totally unaware of one another's presence, and act independently.   There is nothing to them anyway.   But real gases are composed of molecules of finite size.   Hence they might be expected not to follow Dalton's law exactly.   However, Regnault tested the matter experimentally with great care, and found that real gases and even vapors when not too dense obeyed the law very closely. Under ordinary circumstances, then, even real gases are nearly all space.   Their molecules must be very small and far apart.

A further confirmation is afforded by the fact that chemical analyses of air show that the composition of the atmosphere is substantially the same all over the globe.   This means that each of the component gases is thoroughly diffused.   The only component that varies considerably is water vapor.   This is constantly being renewed by evaporation and removed by condensation, and these processes vary considerably from place to place and from time to time, as our variable weather attests. However, the water vapor content seldom exceeds one per cent of the whole, and all of it is confined to within a few miles of the earth. The upper strata are quite dry.

As a result of these facts, the presence of air or of other gases above a liquid can have but little restraining effect upon evaporation. The pressure of a gas on the surface of a liquid does indeed affect the boiling point, because, for a bubble of vapor to form in the interior of the liquid, it must push the liquid away against the opposing pressure, which is substantially that of the gas pressing upon its surface. But evaporation takes place molecule by molecule. When air is at a pressure of one atmosphere and 32 degrees Fahrenheit, only about one five-thousandth part of its volume is occupied by molecules. Hence the molecules of an evaporating liquid shoot into this space almost as freely, and diffuse through it almost as readily, as though there were nothing there. A liquid exposed to the air therefore evaporates nearly as rapidly as in a vacuum.

XIV

Convection and Conduction

Immediately above a hot body, such as a stove, we find a continual stream of warm air ascending. The air in contact with the stove is heated and expands, and becoming thus less dense than the surrounding air, it rises. Cold air is drawn up from below to replace the hot air, and so a continual stream is maintained. This process is called *convection*. It maintains the draught in our fireplaces and chimneys, the ebullition of boiling liquids, the circulation of cooling water in an automobile when there is no pump, etc. Convection carries away great quantities of heat. Moderately hot bodies exposed to the air cool more by this means than by any other.

The ancients were much impressed by the upward leaping of flames. They supposed fire to be one of the four elements, and that to rise was inherent in its nature. Hence the heavens were its natural abode, and all celestial bodies were of a fiery nature. Even Bacon reëchoed this view when, in his *Novum Organum*, he pronounced heat to be an expansive upward motion.

But a flame is only a gas heated to visibility, or containing incandescent carbon particles. It rises for the same reason that any other hot gas rises. It is, in short, a visible convection current, unusually strong, because of the high temperature. But the flame shows only a small part of the whole affair. If we place a lighted candle in the beam of a projection lantern and cast its shadow on a screen, the appearance shown in Figure 40 is produced. The flame itself occupies only the central part of a huge commotion, which looks like a much magnified flame that envelops not only the wick but a good part of the candle as well. This is simply a further portion of the convection current, now rendered visible because it refracts the light differently than the surrounding air does.

By convection, heat is borne away by the steaming of the heated matter itself.   It is possible only in a mobile fluid.   By *conduction*, heat is conveyed from one part to another of a body without any progressive motion of its parts.   It is handed on, so to speak, from molecule to molecule, and spreads like bad news in a crowd.

If a small part of a large body is heated, the molecules in this part are thermally agitated.   Knocking against the surrounding molecules, they communicate their agitation to these, these

FIG. 40—THE SHADOW OF A FLAME

to others, and so on.   The commotion, being thus distributed always among larger and larger groups, weakens as it spreads. The process is slow, and is much like the diffusion of a dissolved substance throughout a solvent.   If only a limited amount of heat is communicated, it finally becomes evenly distributed throughout the body.   The temperature becomes uniform.   If heat is continually supplied, until the temperature of the body rises above that of the surroundings, the molecules at the surface communicate their excess agitation to the neighboring molecules of the air, so that heat escapes from all parts of the surface of the body.   Thus a continuous flow of heat into and out of the body is maintained.

Heat, indeed, can be made to flow along a metallic bar much like water through a pipe. The longer and thinner the bar, the more difficult the passage, just as in the case of a water pipe. For this reason, the handles of our pots and pans are often made of coiled wire, instead of a solid piece. The path that the heat must tread is then long and narrow, and the surface from which it can escape is much increased. But the material of which the bar is composed also makes a difference. Some substances are good conductors, others are poor. Some are so bad that we call them nonconductors or insulators, but there are no perfect insulators.

The metals in general are good heat conductors, the non-metals poor ones. If the pure metals are arranged in the order of their heat conductivities, this will also be the order of their electrical conductivities. There is a close connection between the two. The heat conductivities of a number of common substances are given in the table below in terms of silver = 1000. Mercury is among the poorest metallic heat conductors, with a conductivity one-fiftieth that of silver. The nonmetals range from one two-hundredth that of silver for ice to 1/20,000th for air.

A number of popular illusions have arisen from a lack of understanding of the rôle played by conductivity. If several objects are placed in a constant temperature enclosure, they will all finally come to the same temperature, that of the enclosure. But to the senses they will seem to be of very different temperatures. If the enclosure was cold, a poor conductor like wool will not feel especially cold. A better conductor like stone will feel decidedly cold, while any metal will seem bitingly cold. Conversely, if the enclosure was hot the wool will feel but slightly warm; wood, glass, and porcelain moderately hot; but any metal blisteringly hot. Doubtless from such experiences as these arose the notion that some bodies are naturally cold, others naturally warm. "Stone cold" is still a common expression, as though stone were always cold. And Bacon in his *Organum* listed wool and "all shaggy substances" as naturally warm. The illusion is due to the fact pointed out by Tait, that our sensations of hot and cold do not directly depend upon the temperatures of bodies, but upon the *rate* at which heat is conveyed to or from the skin. A good conductor when cold abstracts heat from the hand faster, and therefore seems colder, than a poor conductor does, even though both have the same temperature. Conversely the good conductor when hot gives up its heat faster, and therefore seems hotter than the poor one. Also, as we shall see presently, the closeness of the contact which the substance makes with the skin affects the rate at which heat passes from one to the other. All "shaggy substances" make poor contact.

It may surprise some to learn that water is a poor conductor of heat, being about 1/770th as good as silver, and only three

Relative Conductivities

Good Conductors

Silver	1000
Copper	940
Aluminum	520
Zinc	270
Tin	159
Iron	155
Lead	85
Mercury	20

Poor Conductors

Ice	5
Glass	1.3-2.5
Porcelain	2.5
Water	1.3
Cotton	0.56
Cardboard	0.5
Hydrogen	0.33
Flannel	0.23
Cork	0.13
Felt	0.09
Cotton wool	0.05
Air	0.05

times better than cotton, as may be seen from the table. The rapid cooling that can be obtained with water is due to convection, not to conduction, and to good contact, as will appear presently. Pure water is also a very poor conductor of electricity. All the liquids are poor conductors of heat, and do not differ much in their conductivities from water. The conductivity of a liquid is difficult to measure, because it is hard to prevent convection currents. These are minimized by heating the liquid at the top.

The poorest conductors of all are the gases. They are still more difficult to measure, not only on account of convection, but because most of the heat is transported directly across them by radiation. Hydrogen and helium are much higher in conductivity than the rest, which are all about the same as air.

One of the best heat insulators is a layer of air, as in the space between double walls. But convection currents must be prevented by filling the space with some loose material like feather stone. The conductivity of cotton by our table is 0.56, but that of cotton wool is 0.05, or about the same as for air. Hence it is

the entrapped air that does all the insulating, not the cotton. The latter serves only to prevent air circulation. And because the conductivity of the cotton itself is higher, we must use as little of it as possible. It must not be packed tightly, but loosely. Similarly fur, felt, and all other "shaggy substances" owe their "warmth" to the entrapped and stagnant air. All heat insulating materials must be light, loose, and porous, in short, contain as little matter as possible.

If the pressure of a gas is very much reduced, its conductivity also falls. A perfect vacuum would be a perfect nonconductor, for there would be no matter present to effect the conduction. We cannot produce a perfect vacuum, but we can come quite near to it.

FIG. 41—THE DEWAR
VACUUM FLASK

The low conductivity of a near vacuum is applied in the Dewar vacuum flask. This is a double walled glass vessel as shown in Figure 41, the space between the double walls being exhausted of air. Convection and conduction are almost wholly absent. A small amount of heat may be conducted up the inner wall and down the outer. But this is a long, narrow, and difficult path, for the glass is thin and a poor conductor. Direct radiation takes place from the inner to the outer wall, and from thence to the air, but this is minimized by silvering the walls, so that some of the radiation is reflected back. This flask was originally invented in 1891 for experiments on liquid air, but is now embodied in the familiar thermos bottle.

Electric light bulbs were originally exhausted because the

carbon filament was combustible.   But the practice is still continued with the modern metallic filaments, because loss of heat by convection and conduction is thereby much reduced.   Since all heat losses must be made good by the electric current, this enables the filament temperature to be maintained with less current, which means a more efficient lamp.

One consequence of the low conductivity of a gas is a phenomenon known as the spheroidal state.   If a drop of water falls on a moderately hot stove, it spreads out and quickly evaporates.   But if the stove is very hot, the water gathers up into a little ball, which runs shimmering and spluttering about, and evaporates much more slowly.   This is the spheroidal state. The hot stove causes such a rapid evaporation at the first instant of contact that the steam blowing downward pushes the drop up out of contact with the stove, and thereafter the drop rests on a cushion of its own steam.   This being a poor conductor of heat prevents further rapid transfer of heat to the drop.   The space between the drop and the stove can be seen with an arrangement like that of Figure 42.   The eye is brought

FIG. 42—THE SPHEROIDAL STATE

to the level of the plate, and one looks at a bright light beyond.

Liquid air has a temperature of -320 degrees Fahrenheit. Compared with that, the hand is a pretty hot stove.   Consequently a drop of liquid air taken on the hand assumes the spheroidal state, and if not allowed to sit in one place too long, will produce no painful impression.   Frozen mercury at -38 degrees Fahrenheit, on the contrary, will produce a severe and painful burn.   One can also dip his hands, if they are moist, into molten iron without injury.   Here it is the steam evaporated from the hand that forms the insulating layer.   Robert Houdin, the famous French magician, had the courage to try this experiment, although his hands of course were invaluable to him in his profession.   It is not, however, an experiment to be recommended.

When one end of a conducting bar, which is heavily insulated as shown in Figure 43, is kept at a constant high temperature, and the other at a constant low temperature, there is a steady flow of heat through the bar; and if the bar is uniform, there is also a steady drop in temperature from the high to the low end, as shown by the dotted line connecting the tops of the mercury columns of the thermometers. The slope of this line, which is given by the total drop in temperature divided by *AB*, is called the *temperature gradient*. It is analogous to the slope of a hill down which water is running. In fact we have here a heat pipe, analogous to a water pipe connecting two reservoirs in

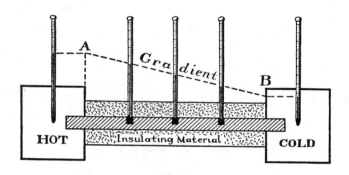

FIG. 43—THE TEMPERATURE GRADIENT

which the water stands at different levels, as in Figure 44. If the levels are kept constant, by pouring water into the one tank, and drawing it off from the other, as fast as the water flows between them, the flow will be steady. And if the pipe is uniform, there will be a steady drop in pressure or *head* along the pipe as shown by the water levels in the open vertical tubes inserted at intervals along the pipe.

Finally, we have the same thing in electricity. When the two ends of a uniform wire are kept at a constant difference of electrical pressure or potential (measured in volts), there is a steady flow of electricity along the wire, and a uniform drop in potential from the high to the low end—a potential gradient.

In all three cases the steadiness of flow depends upon the constancy of the end conditions; the steepness of the gradient depends upon the *resistance* of the bar, pipe, or wire, and the latter upon its cross-section and specific conductivity; the uniformity of the gradient depends upon the uniformity of the bar, pipe, or wire. Wherever there is a constriction, there will be a greater resistance, and the gradient will be steeper. A more rapid fall in temperature, height, or potential will be required to force the heat, water, or electricity through this difficult place.

Especially difficult places occur where two different pieces or bodies come into contact. In electricity we call it contact resistance. Everybody knows that the contact must be good. The abutting surfaces must be large, and they must be clamped, screwed, or otherwise tightly pressed together. If the contact is

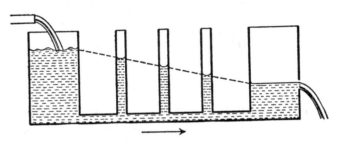

FIG. 44—PRESSURE GRADIENT IN A WATER PIPE

poor, there will be a large drop in potential or temperature at the place, as shown in Figure 45. Solids make poor contact because their surfaces are always rough, as exaggeratedly indicated in the figure, even when they seem quite smooth. Liquids make good contacts with each other and with solids, because they flow into every depression. Hence soldered or welded joints are good, both in electricity and in heat.

A gas and a solid make exceedingly poor contact, for the reason that a gas is mostly space. At any given instant only one three-hundredth to one five-hundredth part of the surface of the solid will be in contact with the gaseous molecules and delivering heat to them by conduction. There is hence great difficulty in transferring heat from a solid to a gas and *vice versa*, especially if it must be done quickly. Large contact surfaces and a great difference in temperature are required. This circumstance has caused the failure of every hot air engine that has ever been devised. It is for this reason that the temperature of a kettle on the stove is much nearer to that of the water it contains, than to that of the fire. The heat of the latter can pass to the kettle only at the expense of a great drop in temperature, 1000 degrees or more. In passing through the metal of the kettle there is almost no drop, for it is a good conductor. In passing from the kettle to the water there is again but little drop, because the contact is excellent. However hot the fire, the kettle is hence at nearly the same temperature as the water. This fact can be strikingly demonstrated by the

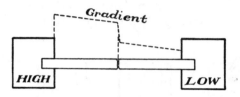

FIG. 45—CONTACT RESISTANCE

little experiment pictured in Figure 46. Water can be boiled in a paper box as shown without even charring the paper. If, however, the flames pass up around the sides of the box, they will burn the paper above the water's edge.

It is for these same reasons that a swimmer gets cold, even if the water is as warm as the air. The water makes close contact with his body and rapidly withdraws heat from it, while the air makes poor contact. If a red-hot poker is exposed to the air, it cools slowly. But if it is plunged into water, it quenches immediately. Liquids are therefore much more effective coolers than gases such as air, not only because of their higher thermal capacities, as already pointed out, but also because of far less contact resistance.

Let us now try another experiment with the heat pipe of Figure 43. Instead of subjecting the two ends to constant temperatures, let us subject the end A to a temperature that alternately rises and falls above and below that to which the end B is subjected, and which is kept constant. Because it takes time for heat to travel, it is evident that at any point along the rod the temperature will not begin to rise until some time after it has started to do so at A. The more distant the point, the greater will be the delay. But if there is no leakage of heat along the way, each point will eventually execute the same alternations of temperature that were impressed at *A*. In short, a *temperature wave* will travel along the rod.

If at some point on the rod, the temperature does not begin to rise until at *A* it has already reached a maximum and descended again to the mean, these two points are said to be in *opposite phase*. For the temperature at the one is rising while that at the other is falling, and *vice versa*. The distance along the bar between the two points is then half a wave length. A

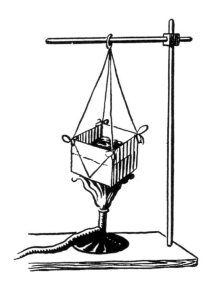

FIG. 46—BOILING WATER IN A PAPER BOX

point at double this distance will be again *in phase* or *in step* with *A*, and this distance will be one wave length.

Since the temperature at *B* rises and falls alternately above and below that of its surroundings, heat will flow out of and into the rod at this point. Instead of a continuous flow, we shall have a back and forth movement of heat in the rod—an alternating current of heat. Thus no heat is permanently transferred from one end to the other. The wave alone is propagated in a definite direction.

We have supposed perfect insulation, but this is never possible. Moreover, there will always be loss from radiation.

Heat will therefore leak out all along the way, and the waves as they advance will diminish in amplitude. Short waves die out more rapidly than long ones. Very short waves have so great a tendency to spread in all directions, that it is impossible to force them very far in any one direction. They die out almost immediately. Since heat travels slowly and is difficult to insulate, only very slow alternations can be transmitted to any considerable distance.

Exactly similar relations hold for electricity. But electricity travels fast, and is easy to insulate. Hence much higher frequencies can be transmitted along a wire, and to much greater distances. But here, too, the higher frequencies die out more rapidly, and a limit in wire transmission is set by the tendency of the short waves to radiate in all directions away from the wire. This tendency of course is made use of in wireless telegraphy.

We are perhaps accustomed to think that these relations apply primarily and particularly to electricity. But they were all discovered and worked out for heat by Fourier in 1822, long before any one dreamed of applying them to electricity.

These temperature waves are exemplified by many common occurrences. Every day at noon the sun reaches its highest point in the heavens, and is then sending us the maximum amount of heat. But the hottest part of the day occurs between two and four in the afternoon, and the coldest part of the night between two and four in the morning. There is hence a lag of from two to four hours in the diurnal temperature wave in the atmosphere near the earth's surface. This is because the sun's heat penetrates the ground a short distance, and warms the objects all about. These take time to cool off, and in so doing continue to warm the air for a time after the sun's rays have begun to decline.

Inside a house there is a further lag on account of the time required for the heat to penetrate the walls, and to be re-emitted by them. Indoors it may still be hot in the evening while outside it is already cool, or cold in the morning when it is already warm outside. If the walls are very thick, as at the base of the Washington Monument, it may be cold in the daytime and warm at night, a lag of twelve hours.

We have also a seasonal wave. The days are longest and we receive most heat in June. They are shortest and we receive the least heat in December.   But the hottest month is August and the coldest is February.   There is here a lag of two months. This is mostly due to the heat that penetrates the ground during the warming, and is slowly given out during the cooling.

We can follow these waves into the ground by placing thermometers at different depths and reading them at intervals. In this way it has been found that the short diurnal variation dies out in from three to four feet; but the long seasonal wave can be followed to a depth of fifty feet or more.   The amplitude diminishes rapidly at first and then more slowly as shown in Figure 47.   Only a few feet down the seasonal wave has so far diminished that the freezing point is never reached.   Pipes laid underground do not freeze, and underground chambers

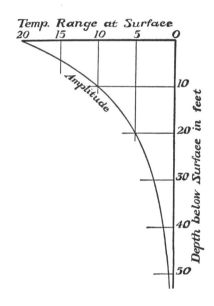

FIG. 47—THE UNDERGROUND SEASONAL TEMPERATURE WAVE

maintain a nearly constant temperature the year round.   There is, of course, also a progressive increase in lag.   At twenty-five feet, the wave is already six months in arrears, so that at this depth it is warm in winter and cool in summer.   The whole variation, however, is then only one-twenty-third of its value at the surface.   At fifty feet, the wave is a year behind, and its amplitude one-five-hundredth of that at the surface—practically no variation at all.

One hot summer's day, when the author was on a hiking trip in Canada, the party came to a spring whose water was extraordinarily cold. The guide told the party that in winter this water was warm. Now this may have been only an effect of contrast; but it may also have been true. If the water came from a depth of twenty-five feet, it would come from a region where the seasons are reversed. However, with a variation at the surface of 92 degrees between summer and winter, there would be at this depth a variation of only 4 degrees. This would scarcely be perceptible to the unaided senses, especially with an interval of six months between the observations to be compared.

XV

RADIATION

If one brings his hand near a hot object, its heat can be felt before the object is touched. Evidently heat is transmitted across the space between the object and the hand. That this can take place without the intervention of matter, is proved by the fact that heat comes to us from the sun across ninety-three million miles of vacuum. It reaches us also from the far more distant stars. In fact, any matter whatsoever in the path of this radiant heat obstructs its passage, and may stop it altogether; for all bodies absorb it more or less, and only a complete vacuum is perfectly transparent to it. By conduction and convection the transfer of heat is effected *by means of matter*, and heat can be transferred in this way only as far as matter extends or can itself be transported. By the process of radiation, however, heat is divorced from its association with matter, and can travel in this way as far as empty space extends, which is to say, indefinitely.

When a total eclipse of the sun occurs, the heat and light received are both cut off at the same instant. Hence both travel in a vacuum with the same speed. Heat rays like light rays travel in straight lines, as is proved by the heat shadows cast by heat opaque objects. Both forms of radiation diminish in intensity as the square of the distance increases. Heat rays can be reflected, refracted, and brought to a focus as by means of a burning glass. In all of these ways they are completely analogous to light rays.

The mechanics of radiation, of its propagation, and absorption, are not yet fully understood. We shall not here discuss the rival wave and quantum theories, the existence or nonexistence of an ether, which are very large questions, but shall confine ourselves to the essential facts, which are independent of any theory.

When a body radiates it is cooled.   The body which receives
and absorbs the radiation is heated.   The transaction therefore represents a transfer of energy.   The ray is born of some
kind of agitation or oscillation.   In the most general sense an
oscillation is merely a repeated series of events that takes place
at a definite locality.   The number of times per second that the
series is repeated, is called the *frequency*.   When a ray is absorbed, it produces in the absorbing body an agitation or
oscillation which also has a frequency.   Hence the ray itself
must embody a frequency.   Experiment shows that with every
ray, at least on its arrival, is associated a very definite frequency,
and that there is a considerable range of these.

If two rays striking at the same point are such that they
produce equal oscillations in opposite phase, they annul each
other's effect.   This is *interference*.   We can obtain it with
sound, adding sound to sound to produce silence.   We can
obtain it with light, adding light to light to produce darkness.
We can obtain it with heat rays, adding heat to heat to produce
absence of heat.

Sound, light, and heat rays will also bend to some extent
around the edges of obstacles.   This is called *diffraction*.   It is
also a frequency effect, for it is large when the frequency is low,
and small when the frequency is high.   It is very large for
sound.   Thus we can hear the shout of a man who is behind a
house, but we cannot see the fire he has built nor feel its warmth.
Nevertheless, by very delicate instruments we can find near the
edge of the shadow both light and heat rays that have bent
slightly into the shadow.   Sound, therefore, has low frequency;
light and heat rays have very high frequencies.

Light and heat we have seen travel at the same speed.   The
speed is very high, 186,000 miles per second.   Sound travels
only eleven hundred feet per second.   Sound also requires the
presence of air or of some other material substance for its propagation, whereas light and heat rays traverse a vacuum
unhindered.   There is yet a third difference.   If we pass light
or heat rays through certain forms of apparatus, we find that it
makes a difference whether a certain part of this apparatus
(which may be a crystal) is rotated on an axis parallel with the
rays, say, from a vertical to a horizontal position.   Interference
will occur in the one position and not in the other.   This effect
is called *polarization*.   It does not occur with sound.

So far, however, we have uncovered no physical distinction between heat and light rays. There remains but one possibility—a difference of frequency. In fact, heat rays are found to have a lower frequency than light rays. This is proved by their greater diffraction or bending into the shadow, or by their less refraction or bending when they enter obliquely a denser medium. When, for example, the sun's rays are brought to a focus by a burning glass, the hottest spot will be found slightly farther from the glass than the brightest spot, showing the less convergence or bending of the heat rays by the lens. But a more satisfactory demonstration is given by a slit and prism arranged as in Figure 48.

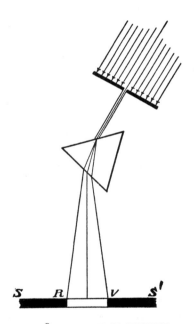

FIG. 48—SPECTRUM OF SUNLIGHT

The white light of the sun contains a number of frequencies. When the light passing the slit falls upon the prism, the rays are differently refracted or bent according to their frequencies, the high frequency rays being bent the more. The rays are thus sorted out according to their frequencies, which to the eye is according to their colors. The white light is thus drawn out into the rainbow band we call the spectrum. Violet light is refracted the most and therefore has the highest frequency. It stands at the end of the spectrum nearer the base of the prism, the right-hand end in the figure. Red is refracted the least, and therefore has the lowest frequency. It stands at the other end of the spectrum. The other colors fall in between. If now we place a series of sensitive thermometers along the spectrum, and some to the right, and some to the left of it, as was first done by William Herschel in 1800, we shall find that they all rise in temperature. But those in the visible part of the spectrum rise very little, those to the right still less. Those to the left, however, show a very considerable rise in temperature. Here, then, in the low frequency region, in the *infra-red*, there are invisible rays that have great heating power. But *all* the rays show *some* heating effect, even those in the *ultra-violet* region to the right. All, then, are really heat rays. Light rays are simply those heat rays that happen to affect the optic nerve of the human being. Nevertheless, for distinction and convenience, and because most of the heat is there, we agree to call the infra-red the real heat spectrum. The extent of the infra-red is very much greater than that of the visible spectrum. The frequency of violet light is about eight hundred million million per second, that of red light is half of this. The infra-red begins, therefore, at a frequency of four hundred million million, and it has been followed down to a frequency of about two million million. In terms of inverse frequencies, which on the wave theory would be wave lengths, the infra-red is two hundred times as long as the visible spectrum. If the latter were one inch in length, the former would be over sixteen feet long. Even this is not the end, for the infra-red merges into the electric spectrum.

The radiation that falls upon a body is partly absorbed, partly transmitted, and partly reflected. If a large proportion is absorbed, the body is *opaque*. If a large proportion is transmitted, the body is *transparent*. A body is not heated by the radiation it transmits or reflects, but only by what it absorbs. Thus, the sunlight that streams through the windowpanes does not heat the glass, but heats the objects in the room that catch and absorb it.

We usually think of bodies as opaque or transparent, according to whether they stop or transmit light. But bodies that are transparent to light are not necessarily transparent to heat rays or *vice versa.* Glass is transparent to both, but water is more opaque to heat. It is for this reason that a water cell is interposed in a projection lantern between the arc and the film to protect the latter. The light rays pass freely, but the heat rays are largely intercepted, and may even bring the water to boil. Fluor spar and rock salt, though opaque to light, are transparent to heat, for which reason prisms and lenses made of these substances are much used in experiments on radiant heat. One must remember also that the infra-red spectrum covers a very large range of frequencies. A body may be transparent to one range and opaque to another, just as is the case with light rays, from which the colours of objects are due. Glass, for example, though transparent to the higher frequency heat rays, is quite opaque to the low frequency ones. Rock salt on the other hand is transparent to the very low frequencies, and is extremely useful on this account.

Many years ago Benjamin Franklin laid pieces of cloth of various colors on snow exposed to the sun, and noted how far each sank by the melting of the snow beneath. The darker ones sank the farther, from which he concluded that they were the better absorbers of heat, and recommended that dark colors be worn in winter and light ones in summer. This conclusion is not entirely justified, however. A black substance absorbs more *light* than a white one, but this tells us nothing about its absorption of *heat.* Tyndall, for example, showed that alum, which is white, is a much better heat absorber than iodine, which is dark; and black amorphous phosphorus is a poor absorber of heat, while sugar is a good one.

Low frequency radiation is associated with low temperature heat, and high frequency with high temperature heat. If we gradually heat a refractory substance, its radiations are at first of very low frequency. But as the heating proceeds, radiations of higher and higher frequencies are added until the body begins to glow with a dull red color. We have at this point reached the low frequency end of the visible spectrum. If the temperature is still further increased, the glow brightens, its color changes to orange, to yellow, and finally to a dazzling white. The whole visible spectrum is now present in addition to the infra-red. It is because the higher the temperature, the greater is the proportion of visible rays to the total heat emitted, that electric lamp filaments are run at the very highest temperatures consistent with a reasonable life of the lamp. The light is then also whiter and more nearly like daylight.

Since all light rays produce heating effect, light entirely without heat is impossible. But the heating effect of the visible rays, especially of those toward the violet end of the spectrum, is exceedingly small. Any means by which these rays could be excited, without at the same time arousing the infra-red, would give a close approximation to light without heat. We have such means in the neon light, in phosphorescence, and in other such phenomena.

The fact that glass is transparent to high frequency heat rays and opaque to low frequency ones, makes it possible to use it as a sort of heat trap. This is the principle of the hothouse. The sunlight enters with its high frequency heat rays and warms the ground and the objects in the hothouse. These re-radiate the heat, but it is now low temperature heat, and the rays are of low frequency. These are stopped by the glass, and so the sun's heat is trapped.

It was suggested by Fourier about 1800 that the earth's atmosphere may similarly act as a heat trap, and that this would account for the comparatively mild and equable climate of our planet in contrast to the violent alternations of temperature that take place on the airless moon. These violent alternations, by the way, were more or less a matter of conjecture in those days, but they have recently been confirmed by measurements made at Mount Wilson, California. During the lunar night, the temperature descends to that of liquid air; during the day, it rises above that of boiling water.

This theory of Fourier's is appropriately called the hothouse theory. It was carefully tested by Tyndall, who in 1859 began a series of measurements of the transparency of gases and vapors to the low temperature radiations of a cube of boiling water. He found, indeed, that these substances were more opaque to the low than to the high frequency radiations, so that there is a hothouse effect. But he also found that some constituents of the atmosphere are much more effective in this respect than others. Water vapor he found to be seventy times, and carbon dioxide ninety times more opaque to the low frequency rays, than dry air. The hothouse effect of the atmosphere therefore depends largely upon the percentages of these constituents present. This doubtless accounts for the more equable climate of humid than of dry regions.

Arrhenius, in his *Worlds in the Making* (1908), drew some rather remarkable conclusions from the high opacity of carbon dioxide. He believes that the warm and equable climate that prevailed all over the earth at the time of the carboniferous age, when coal was laid down in Alaska, Spitzbergen, and even on the Antarctic Continent, was due to the high percentage of carbon dioxide then in the air. He further believes that nearly all of the oxygen now in our atmosphere was produced by the stupendous vegetation of that age, which decomposed the carbon dioxide, depositing the carbon in the ground and liberating the oxygen. Hence, when all our coal is burnt, so too will nearly all the oxygen in the air be exhausted. However, by this replacement of the carbon dioxide in the atmosphere, the earth's climate should again become mild. This may compensate to some extent for the depletion of our coal and oxygen, provided that we can meanwhile develop lung capacity adapted to the stuffy atmosphere. At the rate that our factory chimneys are now pouring carbon dioxide into the air, Arrhenius thinks that a century or two should already make a noticeable difference in the climate. But if man is not ultimately to die of suffocation, or return to the primitive, it is obvious that he must find some other source of mechanical power for his industries than the burning of fuel.

Bodies that are transparent to heat rays are poor conductors of heat. One must not confuse transmission and conduction. All the metals are very opaque to heat rays, because they are good conductors of heat. If we wish to shield ourselves from the radiation of a fire, we should interpose a sheet of metal. Wood, glass, or any other poor conductor of equal thickness, would be much less effective. We have the same relations in electricity. Good conductors of electricity are opaque to electric waves, while insulators are transparent. A thin copper screen will stop radio waves completely, while thick stone or brick walls offer almost no hindrance. In fact the whole radio receiver is operated by the waves that are stopped by the antenna wire.

The rate at which a body emits heat depends upon the nature and extent of the surface, and its temperature. A rumpled or rough surface radiates better than a smooth one.

Hence we provide the cylinders of an air-cooled motor with many pallets and vanes, so as to get as much surface as possible. On the other hand, the calorimeters used in heat measurements are nickel plated and highly polished, so as to reduce the radiating surface. For the same reason the inner walls of the vacuum space of a thermos bottle are coated with quicksilver, since this reduces the radiation as well as reflecting some of it back.

But more than anything else the rate of radiation depends upon the surface temperature. Newton found that for a moderately warm body exposed to the air, the rate of cooling is proportional to the elevation of its temperature above that of the surroundings. This is called Newton's law of cooling. It can easily be verified by filling a flask with hot water, inserting a

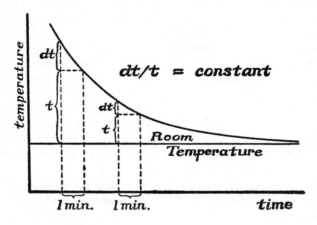

FIG. 49—NEWTON'S COOLING CURVE

thermometer, and reading the latter at intervals. Then making a temperature-time plot, the curve shown in Figure 49 is obtained. At every point of this curve, the ratio of the drop in temperature per minute dt to the temperature *height* t of the point above the room temperature is a constant. The *rate* of cooling at every instant is therefore proportional to the temperature height. It is rapid at first, becomes slower and slower, and approaches the room temperature asymptotically. An infinite time would theoretically be required to cool completely to the room temperature.

If a waiter brings you a cup of black coffee, and you intend

to use sugar and cream, you should put them in at once, instead
of waiting until you are ready to drink the coffee.   This will
lower the temperature at once, but the subsequent loss of heat
by cooling will be less because of this lowered temperature, and
the coffee will be hotter in the end than it would be if you had
delayed putting in the sugar and cream.

Some of our housewives are of the opinion that tea will keep
hot in a silver teapot longer than it will in a china one.   This
they prove to themselves by *feeling* of the pot, which they find
after a lapse of time still feels quite hot, whereas the china pot
does not.   But, as pointed out in the preceding chapter, silver
is an exceedingly good conductor of heat, whereas china is not.
So even if both were at the same temperature, the silver pot
would *feel* hotter.   Furthermore, the good woman feels the
outside of the pot.   Now, because of its high conductivity, the
outside of the silver pot will be nearly as hot as the tea within,
while the outside of the china pot, because of its low conductivity, will be much cooler.   Precisely because of its high
outside temperature, the metal pot cools faster.   The very fact
that it *feels* hot, shows that it parts with its heat readily.   A
thermos bottle, even when filled with a boiling liquid, feels
scarcely warm on the outside.   It is because of this low surface
temperature that it cools slowly.   The whole temperature
gradient is inside the bottle instead of outside.   And you will
not discover after an hour's time if the liquid within is still hot,
by feeling the outside of the bottle.

The curve representing Newton's law of cooling is useful
in many other ways.   Mathematically it is a logarithmic
or exponential curve.   It is the plot of a geometrical against an
arithmetical series.   It is the relation between multiplying and
adding.   It represents the growth or decline of any quantity
where the *rate* is always proportional to the amount on hand.
Hence it represents the growth of capital when put out at compound interest, the increase of population according to the law
of Malthus, the growth of any organism by the multiplication
of its cells,   It represents the rate of radium decay; it is the so-
called logarithmic decrement of damped vibrations, especially
applicable to radio waves.   In Figure 1 (page 4) this same
curve, turned a quarter turn to the right so that the asymptote
becomes vertical, showed the relation between a sensation and
its stimulus known as Weber's law.   In Figure 47, the same

curve, reversed as in a mirror, showed the manner in which temperature waves die out as they penetrate underground. These are only a few of the many applications of this curve, which are found in such diverse fields as physics, economics, sociology, biology, and psychology.

Unfortunately this simple cooling law holds only for moderately warm bodies. Moreover, it is not a pure radiation law, for most of the heat is carried away by convection and some by conduction. If these are prevented or reduced to a very low figure, as by suspending the hot body in a vacuum, it is found that the rate of radiation increases much faster with the temperature than according to Newton's law. Stefan found empirically in 1879 that the rate of radiation is proportional to the fourth power of the absolute temperature of the radiating body, and Boltzman later deduced the same law from theoretical considerations. According to this law, if the absolute temperature of a body is doubled, it radiates sixteen times as fast. At moderate temperatures, then, radiation plays but little part. Most of the heat is carried away by convection and conduction. But at high temperatures it plays the dominant rôle, and in the case of the sun and the stars, which are entirely immersed in a vacuum, it plays the entire rôle. However, Newton's cooling law is sufficiently exact for many purposes, when the bodies concerned are only slightly warmer than their surroundings. It is much used to correct calorimeters and other heat apparatus for loss of heat during the experiment.

It may be noted that Newton's law states that the rate of cooling depends upon the *difference* between the temperature of the body and that of its surroundings, whereas Stefan's law states that the rate of radiation depends upon the absolute temperature of the radiating body alone. According to Stefan, then, a body is always radiating, even if its temperature is much below that of the surroundings. How is it, then, that a body does not cool itself right down to the absolute zero? The answer is given by Prévost's *theory of exchanges*. The cold body radiates to the surroundings, but the surroundings also radiate to the body, and much faster because of their higher temperature. Hence the body receives more heat than it loses. Its temperature rises, while that of the surroundings falls, until the two are equal. Both then radiate at the same rate.

Consider two bodies *A* and *B*, Figure 50, in an enclosure

whose walls are perfectly reflecting. Then all the radiation of each body falls eventually either upon itself or upon the other body, and both will finally assume the same temperature. Let us suppose that A was originally at a white heat, B at a red heat. Then after equalization both may be at a yellow heat. Are we to suppose, that now that the two are at the same temperature, they have ceased to radiate, and that if we could peek into the enclosure, as through a small hole, without disturbing the arrangements, we would see nothing? By no means! We would see two glowing yellow bodies. We are forced then to conclude that at equal temperatures bodies continue to radiate, and therefore that they radiate at all temperatures.

When a body is exposed to radiation, as to that of the sun, its temperature rises until the rate at which it loses heat, by re-radiation and other means, is equal to the rate at which it receives heat. Hence a thermometer set out in the sun will rise above the temperature of the surroundings; but the extent of this rise depends not only upon the intensity of the radiation, but upon the absorbing and reflecting power of the thermometer. It will be quite different with different instruments.

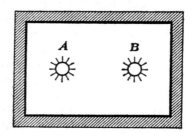

FIG. 50—THE THEORY OF EXCHANGES

Even the same instrument exposed to the same radiation will show a higher temperature, if the reflecting power of the bulb is diminished and its absorbing power increased, by simply blackening it. But the object of a weather thermometer is to indicate the air temperature. Hence it should be used in the shade, and furthermore the bulb should be protected even from such radiations as are present in the shade by a perforated metallic screen over it. Readings taken in the sun have no meaning, and there is no sense to assertions about how hot it is in the sun.

While the rate at which a body radiates depends only upon the nature and temperature of its surface, the rate at which heat can be brought up to the surface from the interior depends upon the conductivity of the body. If it is a poor conductor, heat can only be brought up slowly. Consequently, the surface will be considerably cooler than the interior, and the rate of radiation will be low. A steep temperature gradient will be required to drive the heat to the surface. That is why a thermos bottle, though filled with hot liquid, is scarcely warm on the outside and cools very slowly.

The earth itself is a sort of thermos bottle. As we penetrate below the surface, we find a steady rise in temperature of about one degree Fahrenheit for every sixty feet. At this rate, the melting points of the rocks, even allowing for the elevation produced by pressure, would be exceeded at a depth of only one hundred miles, and at the earth's center the temperature would reach 350,000 degrees Fahrenheit, forty times that of the sun's surface. Hence it was formerly supposed that the earth's interior was molten. But in that case there would be tides in this fluid, like those in the ocean, only stronger. There are earth tides, but they are extremely minute. For these and other reasons, astronomers have come to the conclusion that the earth is solid throughout —except for isolated pockets where volcanoes occur—and very rigid. It is impossible, therefore, that the temperature gradient found near the surface should continue with undiminished steepness to the center. And there are other reasons why it cannot. The average density of the surface rocks is 3.5, that of the earth as a whole is 5.5. Hence the density at the center must be greater than at the surface, probably twice as great, which is the density of iron. And since the earth is magnetic, it is believed that the core *is* mostly iron. Hence as we descend, the materials grow heavier, their conductivity increases, and the temperature gradient must decrease. The interior is hot, but probably not above the melting point of iron at the pressure to which it is there subjected. Thus we have a hot, good conducting core, surrounded by a layer of poorly conducting material analogous to a thermos bottle.

In 1862 Lord Kelvin undertook to determine how long it had taken the earth to cool down from an originally completely molten condition. In his famous paper, "On the Secular Cooling of the Earth," he gave the time as probably one hundred million years; but on account of great uncertainties in the data, he set the rather wide limits of, not more than four hundred million nor less than twenty million years. The calculation was often repeated during the succeeding years, by

Kelvin himself and by others, and they came to favor strongly the lower limit. Tait even reduced it to ten million years. But all this greatly dissatisfied the geologists and biologists, who insistently demanded more time for their evolutionary processes. Besides they had their own methods of reckoning the age of the earth, from the rates of denudation, sedimentation, accumulation of salt in the ocean, etc., all of which gave much higher figures. The physicists looked upon these estimates rather contemptuously, and regarded their own method as the only exact and authoritative one. The discovery of radium in 1898, however, put an entirely different complexion upon the matter. This and other radioactive substances are continually giving off heat, and while they occur only in extremely minute quantities, yet in the whole crust there is enough of them to supply all the heat which comes up from the interior of the earth. The earth at present is apparently not cooling at all, and this balance has probably existed for a long time past. This is in accord with the geological evidence, which does not indicate a hotter and hotter earth as we go back in time, but a very long continuance of temperature conditions much like the present.

While radioactivity has thus invalidated all the old estimates based on secular cooling, it has fortunately provided a new method of measuring geologic time, perhaps more accurate than any other. Uranium disintegrates through a long series of intermediate products, including radium, to lead of a particular atomic weight. From the ratio of the amount of this sort of lead to the uranium in the igneous rocks in which they occur, the age of these rocks can be calculated. These estimates run to one thousand million years and more. Hence the claims of the geologists have been more than justified, and the physicists are now as generous with their allowances to time as they were previously niggardly, and everybody is happy.

BIBLIOGRAPHY

ARRHENIUS, Svante.—*Theories of Chemistry* (Longmans, Green and Company, 1907). Theories concerning hydrates in solutions, Dalton's law, electrolytic conduction.
—— *Worlds in the Making* (Harper and Brothers, 1908).

BRAGG, Sir William.—*Creative Knowledge* (Harper and Brothers, 1927). Chap. II : "The Trade of the Smith."

BRIDGMAN, P. W.—*The Physics of High Pressure* (The Macmillan Company, 1931).

CAJORI, Florian.—*A History of Physics in Its Elementary Branches* (revised and enlarged edition, The Macmillan Company, 1929).

EDSER, Edwin.—*Heat for Advanced Students* (Macmillan and Company, London, 1915).

FOURIER, Joseph.—*The Analytical Theory of Heat*, translated by Alexander Freeman (Dover reprint, 1955).

GUERICKE, Otto von.—*Experimenta Nova Magdeburgica De Vacuo Spatio*, 1672 (German translation in Oswald's Klassiker der exakten Wissenschaften, No. 59, 1894).

JONES, Harry C.—*The Modern Theory of Solution* (Harper and Brothers, 1899). Contains the original memoirs of Pfeffer, Van't Hoff, Arrhenius, and Raoult.

JONES, Harry C.—*The Nature of Solution* (D. Van Nostrand Company, 1917).

MACH, Ernst.—*Principien der Wärmelehre* (Leipzig, 1896).

MAXWELL, J. Clerk.—*Theory of Heat* (tenth edition, Longmans, Green and Company, 1921).

NERNST, Walther.—*Theoretical Chemistry* (revised from the 8th-10th German editions, by L. W. Codd, The Macmillan Company, 1923).
—— *The New Heat Theorem, Its Foundations in Theory and Experiment*, translated from the second German edition by Guy Barr (E. P. Dutton and Company, 1926).

PORTER, Alfred W.—*Thermodynamics* (E. P. Dutton and Company, 1931).

PORTEVIN, A.—"Crystals, Solids and Vitreous Matter," *Scientific American Monthly*, August, 1921 ; translated from *La Revue de l'Ingénieur*, Paris, April, 1921.

POYNTING, J. H., and THOMSON, Sir J. J.—*A Text-Book of Physics, Heat* (ninth edition, G. B. Lippincott Company, 1928).

SHEARCROFT, W. F. F.—*Elementary Heat* (Oxford University Press, 1930).

TAIT, P. G.—*Heat* (Macmillan and Company, London, 1895).

TREVOR, J. E.—*The General Theory of Thermodynamics* (Ginn and Company, 1927).

TYNDALL, John.—*Fragments of Science* (D. Appleton and Company, 1897). Vol. I : chapters on Radiation and on Radiant Heat.

INDEX

CATALOGUE OF DOVER BOOKS

The more difficult books are indicated by an asterisk (*)

Books Explaining Science and Mathematics

WHAT IS SCIENCE?, N. Campbell. The role of experiment and measurement, the function of mathematics, the nature of scientific laws, the difference between laws and theories, the limitations of science, and many similarly provocative topics are treated clearly and without technicalities by an eminent scientist. "Still an excellent introduction to scientific philosophy," H. Margenau in PHYSICS TODAY. "A first-rate primer . . . deserves a wide audience," SCIENTIFIC AMERICAN. 192pp. 5⅜ x 8. S43 Paperbound **$1.25**

THE NATURE OF PHYSICAL THEORY, P. W. Bridgman. A Nobel Laureate's clear, non-technical lectures on difficulties and paradoxes connected with frontier research on the physical sciences. Concerned with such central concepts as thought, logic, mathematics, relativity, probability, wave mechanics, etc. he analyzes the contributions of such men as Newton, Einstein, Bohr, Heisenberg, and many others. "Lucid and entertaining . . . recommended to anyone who wants to get some insight into current philosophies of science," THE NEW PHILOSOPHY. Index. xi + 138pp. 5⅜ x 8. S33 Paperbound **$1.25**

EXPERIMENT AND THEORY IN PHYSICS, Max Born. A Nobel Laureate examines the nature of experiment and theory in theoretical physics and analyzes the advances made by the great physicists of our day: Heisenberg, Einstein, Bohr, Planck, Dirac, and others. The actual process of creation is detailed step-by-step by one who participated. A fine examination of the scientific method at work. 44pp. 5⅜ x 8. S308 Paperbound **75¢**

THE PSYCHOLOGY OF INVENTION IN THE MATHEMATICAL FIELD, J. Hadamard. The reports of such men as Descartes, Pascal, Einstein, Poincaré, and others are considered in this investigation of the method of idea-creation in mathematics and other sciences and the thinking process in general. How do ideas originate? What is the role of the unconscious? What is Poincaré's forgetting hypothesis? are some of the fascinating questions treated. A penetrating analysis of Einstein's thought processes concludes the book. xiii + 145pp. 5⅜ x 8. T107 Paperbound **$1.25**

THE NATURE OF LIGHT AND COLOUR IN THE OPEN AIR, M. Minnaert. Why are shadows sometimes blue, sometimes green, or other colors depending on the light and surroundings? What causes mirages? Why do multiple suns and moons appear in the sky? Professor Minnaert explains these unusual phenomena and hundreds of others in simple, easy-to-understand terms based on optical laws and the properties of light and color. No mathematics is required but artists, scientists, students, and everyone fascinated by these "tricks" of nature will find thousands of useful and amazing pieces of information. Hundreds of observational experiments are suggested which require no special equipment. 200 illustrations; 42 photos. xvi + 362pp. 5⅜ x 8. T196 Paperbound **$2.00**

*MATHEMATICS IN ACTION, O. G. Sutton. Everyone with a command of high school algebra will find this book one of the finest possible introductions to the application of mathematics to physical theory. Ballistics, numerical analysis, waves and wavelike phenomena, Fourier series, group concepts, fluid flow and aerodynamics, statistical measures, and meteorology are discussed with unusual clarity. Some calculus and differential equations theory is developed by the author for the reader's help in the more difficult sections. 88 figures. Index. viii + 236pp. 5⅜ x 8. T440 Clothbound **$3.50**

SOAP-BUBBLES: THEIR COLOURS AND THE FORCES THAT MOULD THEM, C. V. Boys. For continuing popularity and validity as scientific primer, few books can match this volume of easily-followed experiments, explanations. Lucid exposition of complexities of liquid films, surface tension and related phenomena, bubbles' reaction to heat, motion, music, magnetic fields. Experiments with capillary attraction, soap bubbles on frames, composite bubbles, liquid cylinders and jets, bubbles other than soap, etc. Wonderful introduction to scientific method, natural laws that have many ramifications in areas of modern physics. Only complete edition in print. New Introduction by S. Z. Lewin, New York University. 83 illustrations; 1 full-page color plate. xii + 190pp. 5⅜ x 8½. T542 Paperbound **95¢**

CATALOGUE OF DOVER BOOKS

THE STORY OF X-RAYS FROM RÖNTGEN TO ISOTOPES, A. R. Bleich, M.D. This book, by a member of the American College of Radiology, gives the scientific explanation of x-rays, their applications in medicine, industry and art, and their danger (and that of atmospheric radiation) to the individual and the species. You learn how radiation therapy is applied against cancer, how x-rays diagnose heart disease and other ailments, how they are used to examine mummies for information on diseases of early societies, and industrial materials for hidden weaknesses. 54 illustrations show x-rays of flowers, bones, stomach, gears with flaws, etc. 1st publication. Index. xix + 186pp. 5⅜ x 8.　　　　　　　　　　T622 Paperbound **$1.35**

SPINNING TOPS AND GYROSCOPIC MOTION, John Perry. A classic elementary text of the dynamics of rotation — the behavior and use of rotating bodies such as gyroscopes and tops. In simple, everyday English you are shown how quasi-rigidity is induced in discs of paper, smoke rings, chains, etc., by rapid motions; why a gyrostat falls and why a top rises; precession; how the earth's motion affects climate; and many other phenomena. Appendix on practical use of gyroscopes. 62 figures. 128pp. 5⅜ x 8.　　　　　T416 Paperbound **$1.00**

SNOW CRYSTALS, W. A. Bentley, M. J. Humphreys. For almost 50 years W. A. Bentley photographed snow flakes in his laboratory in Jericho, Vermont; in 1931 the American Meteorological Society gathered together the best of his work, some 2400 photographs of snow flakes, plus a few ice flowers, windowpane frosts, dew, frozen rain, and other ice formations. Pictures were selected for beauty and scientific value. A very valuable work to anyone in meteorology, cryology; most interesting to layman; extremely useful for artist who wants beautiful, crystalline designs. All copyright free. Unabridged reprint of 1931 edition. 2453 illustrations. 227pp. 8 x 10½.　　　　　　　　　　　　T287 Paperbound **$3.00**

A DOVER SCIENCE SAMPLER, edited by George Barkin. A collection of brief, non-technical passages from 44 Dover Books Explaining Science for the enjoyment of the science-minded browser. Includes work of Bertrand Russell, Poincaré, Laplace, Max Born, Galileo, Newton; material on physics, mathematics, metallurgy, anatomy, astronomy, chemistry, etc. You will be fascinated by Martin Gardner's analysis of the sincere pseudo-scientist, Moritz's account of Newton's absentmindedness, Bernard's examples of human vivisection, etc. Illustrations from the Diderot Pictorial Encyclopedia and De Re Metallica. 64 pages.　　**FREE**

THE STORY OF ATOMIC THEORY AND ATOMIC ENERGY, J. G. Feinberg. A broader approach to subject of nuclear energy and its cultural implications than any other similar source. Very readable, informal, completely non-technical text. Begins with first atomic theory, 600 B.C. and carries you through the work of Mendelejeff, Röntgen, Madame Curie, to Einstein's equation and the A-bomb. New chapter goes through thermonuclear fission, binding energy, other events up to 1959. Radioactive decay and radiation hazards, future benefits, work of Bohr, moderns, hundreds more topics. "Deserves special mention . . . not only authoritative but thoroughly popular in the best sense of the word," Saturday Review. Formerly, "The Atom Story." Expanded with new chapter. Three appendixes. Index. 34 illustrations. vii + 243pp. 5⅜ x 8.　　　　　　　　　　　　　T625 Paperbound **$1.60**

THE STRANGE STORY OF THE QUANTUM, AN ACCOUNT FOR THE GENERAL READER OF THE GROWTH OF IDEAS UNDERLYING OUR PRESENT ATOMIC KNOWLEDGE, B. Hoffmann. Presents lucidly and expertly, with barest amount of mathematics, the problems and theories which led to modern quantum physics. Dr. Hoffmann begins with the closing years of the 19th century, when certain trifling discrepancies were noticed, and with illuminating analogies and examples takes you through the brilliant concepts of Planck, Einstein, Pauli, Broglie, Bohr, Schroedinger, Heisenberg, Dirac, Sommerfeld, Feynman, etc. This edition includes a new, long postscript carrying the story through 1958. "Of the books attempting an account of the history and contents of our modern atomic physics which have come to my attention, this is the best," H. Margenau, Yale University, in "American Journal of Physics." 32 tables and line illustrations. Index. 275pp. 5⅜ x 8.　　　　　　　　　T518 Paperbound **$1.50**

SPACE AND TIME, E. Borel. Written by a versatile mathematician of world renown with his customary lucidity and precision, this introduction to relativity for the layman presents scores of examples, analogies, and illustrations that open up new ways of thinking about space and time. It covers abstract geometry and geographical maps, continuity and topology, the propagation of light, the special theory of relativity, the general theory of relativity, theoretical researches, and much more. Mathematical notes. 2 Indexes. 4 Appendices. 15 figures. xvi + 243pp. 5⅜ x 8.　　　　　　　　　　　　　T592 Paperbound **$1.45**

FROM EUCLID TO EDDINGTON: A STUDY OF THE CONCEPTIONS OF THE EXTERNAL WORLD, Sir Edmund Whittaker. A foremost British scientist traces the development of theories of natural philosophy from the western rediscovery of Euclid to Eddington, Einstein, Dirac, etc. The inadequacy of classical physics is contrasted with present day attempts to understand the physical world through relativity, non-Euclidean geometry, space curvature, wave mechanics, etc. 5 major divisions of examination: Space; Time and Movement; the Concepts of Classical Physics; the Concepts of Quantum Mechanics; the Eddington Universe. 212pp. 5⅜ x 8. 　　　　　　　　　　　　　　　　　　　　　　　　T491 Paperbound **$1.35**

CATALOGUE OF DOVER BOOKS

*THE EVOLUTION OF SCIENTIFIC THOUGHT FROM NEWTON TO EINSTEIN, A. d'Abro. A detailed account of the evolution of classical physics into modern relativistic theory and the concommitant changes in scientific methodology. The breakdown of classical physics in the face of non-Euclidean geometry and the electromagnetic equations is carefully discussed and then an exhaustive analysis of Einstein's special and general theories of relativity and their implications is given. Newton, Riemann, Weyl, Lorentz, Planck, Maxwell, and many others are considered. A non-technical explanation of space, time, electromagnetic waves, etc. as understood today. "Model of semi-popular exposition," NEW REPUBLIC. 21 diagrams. 482pp. 5⅜ x 8.
T2 Paperbound **$2.25**

EINSTEIN'S THEORY OF RELATIVITY, Max Born. Nobel Laureate explains Einstein's special and general theories of relativity, beginning with a thorough review of classical physics in simple, non-technical language. Exposition of Einstein's work discusses concept of simultaneity, kinematics, relativity of arbitrary motions, the space-time continuum, geometry of curved surfaces, etc., steering middle course between vague popularizations and complex scientific presentations. 1962 edition revised by author takes into account latest findings, predictions of theory and implications for cosmology, indicates what is being sought in unified field theory. Mathematics very elementary, illustrative diagrams and experiments informative but simple. Revised 1962 edition. Revised by Max Born, assisted by Gunther Leibfried and Walter Biem. Index. 143 illustrations. vii + 376pp. 5⅜ x 8.
S769 Paperbound **$2.00**

PHILOSOPHY AND THE PHYSICISTS, L. Susan Stebbing. A philosopher examines the philosophical aspects of modern science, in terms of a lively critical attack on the ideas of Jeans and Eddington. Such basic questions are treated as the task of science, causality, determinism, probability, consciousness, the relation of the world of physics to the world of everyday experience. The author probes the concepts of man's smallness before an inscrutable universe, the tendency to idealize mathematical construction, unpredictability theorems and human freedom, the supposed opposition between 19th century determinism and modern science, and many others. Introduces many thought-stimulating ideas about the implications of modern physical concepts. xvi + 295pp. 5⅜ x 8.
T480 Paperbound **$1.65**

THE RESTLESS UNIVERSE, Max Born. A remarkably lucid account by a Nobel Laureate of recent theories of wave mechanics, behavior of gases, electrons and ions, waves and particles, electronic structure of the atom, nuclear physics, and similar topics. "Much more thorough and deeper than most attempts . . . easy and delightful," CHEMICAL AND ENGINEERING NEWS. Special feature: 7 animated sequences of 60 figures each showing such phenomena as gas molecules in motion, the scattering of alpha particles, etc. 11 full-page plates of photographs. Total of nearly 600 illustrations. 351pp. 6⅛ x 9¼.
T412 Paperbound **$2.00**

THE COMMON SENSE OF THE EXACT SCIENCES, W. K. Clifford. For 70 years a guide to the basic concepts of scientific and mathematical thought. Acclaimed by scientists and laymen alike, it offers a wonderful insight into concepts such as the extension of meaning of symbols, characteristics of surface boundaries, properties of plane figures, measurement of quantities, vectors, the nature of position, bending of space, motion, mass and force, and many others. Prefaces by Bertrand Russell and Karl Pearson. Critical introduction by James Newman. 130 figures. 249pp. 5⅜ x 8.
T61 Paperbound **$1.60**

MATTER AND LIGHT, THE NEW PHYSICS, Louis de Broglie. Non-technical explanations by a Nobel Laureate of electro-magnetic theory, relativity, matter, light and radiation, wave mechanics, quantum physics, philosophy of science, and similar topics. This is one of the simplest yet most accurate introductions to the work of men like Planck, Einstein, Bohr, and others. Only 2 of the 21 chapters require a knowledge of mathematics. 300pp. 5⅜ x 8.
T35 Paperbound **$1.85**

SCIENCE, THEORY AND MAN, Erwin Schrödinger. This is a complete and unabridged reissue of SCIENCE AND THE HUMAN TEMPERAMENT plus an additional essay: "What Is an Elementary Particle?" Nobel Laureate Schrödinger discusses such topics as nature of scientific method, the nature of science, chance and determinism, science and society, conceptual models for physical entities, elementary particles and wave mechanics. Presentation is popular and may be followed by most people with little or no scientific training. "Fine practical preparation for a time when laws of nature, human institutions . . . are undergoing a critical examination without parallel," Waldemar Kaempffert, N. Y. TIMES. 192pp. 5⅜ x 8.
T428 Paperbound **$1.35**

CONCERNING THE NATURE OF THINGS, Sir William Bragg. The Nobel Laureate physicist in his Royal Institute Christmas Lectures explains such diverse phenomena as the formation of crystals, how uranium is transmuted to lead, the way X-rays work, why a spinning ball travels in a curved path, the reason why bubbles bounce from each other, and many other scientific topics that are seldom explained in simple terms. No scientific background needed—book is easy enough that any intelligent adult or youngster can understand it. Unabridged. 32pp. of photos; 57 figures. xii + 232pp. 5⅜ x 8.
T31 Paperbound **$1.35**

*THE RISE OF THE NEW PHYSICS (formerly THE DECLINE OF MECHANISM), A. d'Abro. This authoritative and comprehensive 2 volume exposition is unique in scientific publishing. Written for intelligent readers not familiar with higher mathematics, it is the only thorough explanation in non-technical language of modern mathematical-physical theory. Combining both history and exposition, it ranges from classical Newtonian concepts up through the electronic theories of Dirac and Heisenberg, the statistical mechanics of Fermi, and Einstein's relativity theories. "A must for anyone doing serious study in the physical sciences," J. OF FRANKLIN INST. 97 illustrations. 991pp. 2 volumes.
T3 Vol. 1, Paperbound **$2.25**
T4 Vol. 2, Paperbound **$2.25**

CATALOGUE OF DOVER BOOKS

SCIENCE AND HYPOTHESIS, Henri Poincaré. Creative psychology in science. How such concepts as number, magnitude, space, force, classical mechanics were developed and how the modern scientist uses them in his thought. Hypothesis in physics, theories of modern physics. Introduction by Sir James Larmor. "Few mathematicians have had the breadth of vision of Poincaré, and none is his superior in the gift of clear exposition," E. T. Bell. Index. 272pp. 5⅜ x 8.
S221 Paperbound **$1.35**

THE VALUE OF SCIENCE, Henri Poincaré. Many of the most mature ideas of the "last scientific universalist" conveyed with charm and vigor for both the beginning student and the advanced worker. Discusses the nature of scientific truth, whether order is innate in the universe or imposed upon it by man, logical thought versus intuition (relating to mathematics through the works of Weierstrass, Lie, Klein, Riemann), time and space (relativity, psychological time, simultaneity), Hertz's concept of force, interrelationship of mathematical physics to pure math, values within disciplines of Maxwell, Carnot, Mayer, Newton, Lorentz, etc. Index. iii + 147pp. 5⅜ x 8.
S469 Paperbound **$1.35**

THE SKY AND ITS MYSTERIES, E. A. Beet. One of the most lucid books on the mysteries of the universe; covers history of astronomy from earliest observations to modern theories of expanding universe, source of stellar energy, birth of planets, origin of moon craters, possibilities of life on other planets. Discusses effects of sunspots on weather; distance, age of stars; methods and tools of astronomers; much more. Expert and fascinating. "Eminently readable book," London Times. Bibliography. Over 50 diagrams, 12 full-page plates. Fold-out star map. Introduction. Index. 238pp. 5¼ x 7½.
T627 Clothbound **$3.50**

OUT OF THE SKY: AN INTRODUCTION TO METEORITICS, H. H. Nininger. A non-technical yet comprehensive introduction to the young science of meteoritics: all aspects of the arrival of cosmic matter on our planet from outer space and the reaction and alteration of this matter in the terrestrial environment. Essential facts and major theories presented by one of the world's leading experts. Covers ancient reports of meteors; modern systematic investigations; fireball clusters; meteorite showers; tektites; planetoidal encounters; etc. 52 full-page plates with over 175 photographs. 22 figures. Bibliography and references. Index. viii + 336pp. 5⅜ x 8.
T519 Paperbound **$1.85**

THE REALM OF THE NEBULAE, E. Hubble. One of great astronomers of our day records his formulation of concept of "island universes." Covers velocity-distance relationship; classification, nature, distances, general types of nebulae; cosmological theories. A fine introduction to modern theories for layman. No math needed. New introduction by A. Sandage. 55 illustrations, photos. Index. iv + 201pp. 5⅜ x 8.
S455 Paperbound **$1.50**

AN ELEMENTARY SURVEY OF CELESTIAL MECHANICS, Y. Ryabov. Elementary exposition of gravitational theory and celestial mechanics. Historical introduction and coverage of basic principles, including: the ecliptic, the orbital plane, the 2- and 3-body problems, the discovery of Neptune, planetary rotation, the length of the day, the shapes of galaxies, satellites (detailed treatment of Sputnik I), etc. First American reprinting of successful Russian popular exposition. Follow actual methods of astrophysicists with only high school math! Appendix. 58 figures. 165pp. 5⅜ x 8.
T756 Paperbound **$1.25**

GREAT IDEAS AND THEORIES OF MODERN COSMOLOGY, Jagjit Singh. Companion volume to author's popular "Great Ideas of Modern Mathematics" (Dover, $1.55). The best nontechnical survey of post-Einstein attempts to answer perhaps unanswerable questions of origin, age of Universe, possibility of life on other worlds, etc. Fundamental theories of cosmology and cosmogony recounted, explained, evaluated in light of most recent data: Einstein's concepts of relativity, space-time; Milne's a priori world-system; astrophysical theories of Jeans, Eddington; Hoyle's "continuous creation;" contributions of dozens more scientists. A faithful, comprehensive critical summary of complex material presented in an extremely well-written text intended for laymen. Original publication. Index. xii + 276pp. 5⅜ x 8½.
T925 Paperbound **$1.85**

BASIC ELECTRICITY, Bureau of Naval Personnel. Very thorough, easily followed course in basic electricity for beginner, layman, or intermediate student. Begins with simplest definitions, presents coordinated, systematic coverage of basic theory and application: conductors, insulators, static electricity, magnetism, production of voltage, Ohm's law, direct current series and parallel circuits, wiring techniques, electromagnetism, alternating current, capacitance and inductance, measuring instruments, etc.; application to electrical machines such as alternating and direct current generators, motors, transformers, magnetic magnifiers, etc. Each chapter contains problems to test progress; answers at rear. No math needed beyond algebra. Appendices on signs, formulas, etc. 345 illustrations. 448pp. 7½ x 10.
S973 Paperbound **$3.00**

ELEMENTARY METALLURGY AND METALLOGRAPHY, A. M. Shrager. An introduction to common metals and alloys; stress is upon steel and iron, but other metals and alloys also covered. All aspects of production, processing, working of metals. Designed for student who wishes to enter metallurgy, for bright high school or college beginner, layman who wants background on extremely important industry. Questions, at ends of chapters, many microphotographs, glossary. Greatly revised 1961 edition. 195 illustrations, tables. ix + 389pp. 5⅜ x 8.
S138 Paperbound **$2.25**

CATALOGUE OF DOVER BOOKS

BRIDGES AND THEIR BUILDERS, D. B. Steinman & S. R. Watson. Engineers, historians, and every person who has ever been fascinated by great spans will find this book an endless source of information and interest. Greek and Roman structures, Medieval bridges, modern classics such as the Brooklyn Bridge, and the latest developments in the science are retold by one of the world's leading authorities on bridge design and construction. BRIDGES AND THEIR BUILDERS is the only comprehensive and accurate semi-popular history of these important measures of progress in print. New, greatly revised, enlarged edition. 23 photos; 26 line-drawings. Index. xvii + 401pp. 5⅜ x 8.   T431 Paperbound **$2.00**

FAMOUS BRIDGES OF THE WORLD, D. B. Steinman. An up-to-the-minute new edition of a book that explains the fascinating drama of how the world's great bridges came to be built. The author, designer of the famed Mackinac bridge, discusses bridges from all periods and all parts of the world, explaining their various types of construction, and describing the problems their builders faced. Although primarily for youngsters, this cannot fail to interest readers of all ages. 48 illustrations in the text. 23 photographs. 99pp. 6⅛ x 9¼.   T161 Paperbound **$1.00**

HOW DO YOU USE A SLIDE RULE? by A. A. Merrill. A step-by-step explanation of the slide rule that presents the fundamental rules clearly enough for the non-mathematician to understand. Unlike most instruction manuals, this work concentrates on the two most important operations: multiplication and division. 10 easy lessons, each with a clear drawing, for the reader who has difficulty following other expositions. 1st publication. Index. 2 Appendices. 10 illustrations. 78 problems, all with answers. vi + 36 pp. 6⅛ x 9¼.   T62 Paperbound **60¢**

HOW TO CALCULATE QUICKLY, H. Sticker. A tried and true method for increasing your "number sense" — the ability to see relationships between numbers and groups of numbers. Addition, subtraction, multiplication, division, fractions, and other topics are treated through techniques not generally taught in schools: left to right multiplication, division by inspection, etc. This is not a collection of tricks which work only on special numbers, but a detailed well-planned course, consisting of over 9,000 problems that you can work in spare moments. It is excellent for anyone who is inconvenienced by slow computational skills. 5 or 10 minutes of this book daily will double or triple your calculation speed. 9,000 problems, answers. 256pp. 5⅜ x 8.   T295 Paperbound **$1.00**

MATHEMATICAL FUN, GAMES AND PUZZLES, Jack Frohlichstein. A valuable service for parents of children who have trouble with math, for teachers in need of a supplement to regular upper elementary and junior high math texts (each section is graded—easy, average, difficult —for ready adaptation to different levels of ability), and for just anyone who would like to develop basic skills in an informal and entertaining manner. The author combines ten years of experience as a junior high school math teacher with a method that uses puzzles and games to introduce the basic ideas and operations of arithmetic. Stress on everyday uses of math: banking, stock market, personal budgets, insurance, taxes. Intellectually stimulating and practical, too. 418 problems and diversions with answers. Bibliography. 120 illustrations. xix + 306pp. 5⅝ x 8½.   T789 Paperbound **$1.75**

GREAT IDEAS OF MODERN MATHEMATICS: THEIR NATURE AND USE, Jagjit Singh. Reader with only high school math will understand main mathematical ideas of modern physics, astronomy, genetics, psychology, evolution, etc. better than many who use them as tools, but comprehend little of their basic structure. Author uses his wide knowledge of non-mathematical fields in brilliant exposition of differential equations, matrices, group theory, logic, statistics, problems of mathematical foundations, imaginary numbers, vectors, etc. Original publication. 2 appendixes. 2 indexes. 65 illustr. 322pp. 5⅜ x 8.   S587 Paperbound **$1.75**

THE UNIVERSE OF LIGHT, W. Bragg. Sir William Bragg, Nobel Laureate and great modern physicist, is also well known for his powers of clear exposition. Here he analyzes all aspects of light for the layman: lenses, reflection, refraction, the optics of vision, x-rays, the photoelectric effect, etc. He tells you what causes the color of spectra, rainbows, and soap bubbles, how magic mirrors work, and much more. Dozens of simple experiments are described. Preface. Index. 199 line drawings and photographs, including 2 full-page color plates. x + 283pp. 5⅜ x 8.   T538 Paperbound **$1.85**

*INTRODUCTION TO SYMBOLIC LOGIC AND ITS APPLICATIONS, Rudolph Carnap. One of the clearest, most comprehensive, and rigorous introductions to modern symbolic logic, by perhaps its greatest living master. Not merely elementary theory, but demonstrated applications in mathematics, physics, and biology. Symbolic languages of various degrees of complexity are analyzed, and one constructed. "A creation of the rank of a masterpiece," Zentralblatt für Mathematik und Ihre Grenzgebiete. Over 300 exercises. 5 figures. Bibliography. Index. xvi + 241pp. 5⅜ x 8.   S453 Paperbound **$1.85**

*HIGHER MATHEMATICS FOR STUDENTS OF CHEMISTRY AND PHYSICS, J. W. Mellor. Not abstract, but practical, drawing its problems from familiar laboratory material, this book covers theory and application of differential calculus, analytic geometry, functions with singularities, integral calculus, infinite series, solution of numerical equations, differential equations, Fourier's theorem and extensions, probability and the theory of errors, calculus of variations, determinants, etc. "If the reader is not familiar with this book, it will repay him to examine it," CHEM. & ENGINEERING NEWS. 800 problems. 189 figures. 2 appendices; 30 tables of integrals, probability functions, etc. Bibliography. xxi + 641pp. 5⅜ x 8.
                                                          S193 Paperbound **$2.50**

THE FOURTH DIMENSION SIMPLY EXPLAINED, edited by Henry P. Manning. ·Originally written as entries in contest sponsored by "Scientific American," then published in book form, these 22 essays present easily understood explanations of how the fourth dimension may be studied, the relationship of non-Euclidean geometry to the fourth dimension, analogies to three-dimensional space, some fourth-dimensional absurdities and curiosities, possible measurements and forms in the fourth dimension. In general, a thorough coverage of many of the simpler properties of fourth-dimensional space. Multi-points of view on many of the most important aspects are valuable aid to comprehension. Introduction by Dr. Henry P. Manning gives proper emphasis to points in essays, more advanced account of fourth-dimensional geometry. 82 figures. 251pp. 5⅜ x 8.   T711 Paperbound **$1.35**

TRIGONOMETRY REFRESHER FOR TECHNICAL MEN, A. A. Klaf. A modern question and answer text on plane and spherical trigonometry. Part I covers plane trigonometry: angles, quadrants, trigonometrical functions, graphical representation, interpolation, equations, logarithms, solution of triangles, slide rules, etc. Part II discusses applications to navigation, surveying, elasticity, architecture, and engineering. Small angles, periodic functions, vectors, polar coordinates, De Moivre's theorem, fully covered. Part III is devoted to spherical trigonometry and the solution of spherical triangles, with applications to terrestrial and astronomical problems. Special time-savers for numerical calculation. 913 questions answered for you! 1738 problems; answers to odd numbers. 494 figures. 14 pages of functions, formulae. Index. x + 629pp. 5⅜ x 8.   T371 Paperbound **$2.00**

CALCULUS REFRESHER FOR TECHNICAL MEN. A. A. Klaf. Not an ordinary textbook but a unique refresher for engineers, technicians, and students. An examination of the most important aspects of differential and integral calculus by means of 756 key questions. Part I covers simple differential calculus: constants, variables, functions, increments, derivatives, logarithms, curvature, etc. Part II treats fundamental concepts of integration: inspection, substitution, transformation, reduction, areas and volumes, mean value, successive and partial integration, double and triple integration. Stresses practical aspects! A 50 page section gives applications to civil and nautical engineering, electricity, stress and strain, elasticity, industrial engineering, and similar fields. 756 questions answered. 556 problems; solutions to odd numbers. 36 pages of constants, formulae. Index. v + 431pp. 5⅜ x 8.

T370 Paperbound **$2.00**

PROBABILITIES AND LIFE, Emile Borel. One of the leading French mathematicians of the last 100 years makes use of certain results of mathematics of probabilities and explains a number of problems that for the most part, are related to everyday living or to illness and death: computation of life expectancy tables, chances of recovery from various diseases, probabilities of job accidents, weather predictions, games of chance, and so on. Emphasis on results not processes, though some indication is made of mathematical proofs. Simple in style, free of technical terminology, limited in scope to everyday situations, it is comprehensible to laymen, fine reading for beginning students of probability. New English translation. Index. Appendix. vi + 87pp. 5⅜ x 8½.   T121 Paperbound **$1.00**

POPULAR SCIENTIFIC LECTURES, Hermann von Helmholtz. 7 lucid expositions by a pre-eminent scientific mind: "The Physiological Causes of Harmony in Music," "On the Relation of Optics to Painting," "On the Conservation of Force," "On the Interaction of Natural Forces," "On Goethe's Scientific Researches" into theory of color, "On the Origin and Significance of Geometric Axioms," "On Recent Progress in the Theory of Vision." Written with simplicity of expression, stripped of technicalities, these are easy to understand and delightful reading for anyone interested in science or looking for an introduction to serious study of acoustics or optics. Introduction by Professor Morris Kline, Director, Division of Electromagnetic Research, New York University, contains astute, impartial evaluations. Selected from "Popular Lectures on Scientific Subjects," 1st and 2nd series. xii + 286pp. 5⅜ x 8½.   T799 Paperbound **$1.45**

SCIENCE AND METHOD, Henri Poincaré. Procedure of scientific discovery, methodology, experiment, idea-germination—the intellectual processes by which discoveries come into being. Most significant and most interesting aspects of development, application of ideas. Chapters cover selection of facts, chance, mathematical reasoning, mathematics, and logic; Whitehead, Russell, Cantor; the new mechanics, etc. 288pp. 5⅜ x 8.   S222 Paperbound **$1.50**

HEAT AND ITS WORKINGS, Morton Mott-Smith, Ph.D. An unusual book; to our knowledge the only middle-level survey of this important area of science. Explains clearly such important concepts as physiological sensation of heat and Weber's law, measurement of heat, evolution of thermometer, nature of heat, expansion and contraction of solids, Boyle's law, specific heat. BTU's and calories, evaporation, Andrews's isothermals, radiation, the relation of heat to light, many more topics inseparable from other aspects of physics. A wide, non-mathematical yet thorough explanation of basic ideas, theories, phenomena for laymen and beginning scientists illustrated by experiences of daily life. Bibliography. 50 illustrations. x + 165pp. 5⅜ x 8½.   T978 Paperbound **$1.00**

Classics of Science

THE DIDEROT PICTORIAL ENCYCLOPEDIA OF TRADES AND INDUSTRY, MANUFACTURING AND THE TECHNICAL ARTS IN PLATES SELECTED FROM "L'ENCYCLOPEDIE OU DICTIONNAIRE RAISONNE DES SCIENCES, DES ARTS, ET DES METIERS" OF DENIS DIDEROT, edited with text by C. Gillispie. The first modern selection of plates from the high point of 18th century French engraving, Diderot's famous Encyclopedia. Over 2000 illustrations on 485 full page plates, most of them original size, illustrating the trades and industries of one of the most fascinating periods of modern history, 18th century France. These magnificent engravings provide an invaluable glimpse into the past for the student of early technology, a lively and accurate social document to students of cultures, an outstanding find to the lover of fine engravings. The plates teem with life, with men, women, and children performing all of the thousands of operations necessary to the trades before and during the early stages of the industrial revolution. Plates are in sequence, and show general operations, closeups of difficult operations, and details of complex machinery. Such important and interesting trades and industries are illustrated as sowing, harvesting, beekeeping, cheesemaking, operating windmills, milling flour, charcoal burning, tobacco processing, indigo, fishing, arts of war, salt extraction, mining, smelting iron, casting iron, steel, extracting mercury, zinc, sulphur, copper, etc., slating, tinning, silverplating, gilding, making gunpowder, cannons, bells, shoeing horses, tanning, papermaking, printing, dying, and more than 40 other categories. 920pp. 9 x 12. Heavy library cloth. T421 Two volume set **$18.50**

THE PRINCIPLES OF SCIENCE, A TREATISE ON LOGIC AND THE SCIENTIFIC METHOD, W. Stanley Jevons. Treating such topics as Inductive and Deductive Logic, the Theory of Number, Probability, and the Limits of Scientific Method, this milestone in the development of symbolic logic remains a stimulating contribution to the investigation of inferential validity in the natural and social sciences. It significantly advances Boole's logic, and describes a machine which is a foundation of modern electronic calculators. In his introduction, Ernest Nagel of Columbia University says, "(Jevons) . . . continues to be of interest as an attempt to articulate the logic of scientific inquiry." Index. liii + 786pp. 5⅜ x 8.
S446 Paperbound **$2.98**

*DIALOGUES CONCERNING TWO NEW SCIENCES, Galileo Galilei. A classic of experimental science which has had a profound and enduring influence on the entire history of mechanics and engineering. Galileo based this, his finest work, on 30 years of experimentation. It offers a fascinating and vivid exposition of dynamics, elasticity, sound, ballistics, strength of materials, and the scientific method. Translated by H. Crew and A. de Salvio. 126 diagrams. Index. xxi + 288pp. 5⅜ x 8. S99 Paperbound **$1.75**

DE MAGNETE, William Gilbert. This classic work on magnetism founded a new science. Gilbert was the first to use the word "electricity," to recognize mass as distinct from weight, to discover the effect of heat on magnetic bodies; invented an electroscope, differentiated between static electricity and magnetism, conceived of the earth as a magnet. Written by the first great experimental scientist, this lively work is valuable not only as an historical landmark, but as the delightfully easy-to-follow record of a perpetually searching, ingenious mind. Translated by P. F. Mottelay. 25 page biographical memoir. 90 fix. lix + 368pp. 5⅜ x 8. S470 Paperbound **$2.00**

*OPTICKS, Sir Isaac Newton. An enormous storehouse of insights and discoveries on light, reflection, color, refraction, theories of wave and corpuscular propagation of light, optical apparatus, and mathematical devices which have recently been reevaluated in terms of modern physics and placed in the top-most ranks of Newton's work! Foreword by Albert Einstein. Preface by I. B. Cohen of Harvard U. 7 pages of portraits, facsimile pages, letters, etc. cxvi + 412pp. 5⅜ x 8. S205 Paperbound **$2.25**

A SURVEY OF PHYSICAL THEORY, M. Planck. Lucid essays on modern physics for the general reader by the Nobel Laureate and creator of the quantum revolution. Planck explains how the new concepts came into being; explores the clash between theories of mechanics, electrodynamics, and thermodynamics; and traces the evolution of the concept of light through Newton, Huygens, Maxwell, and his own quantum theory, providing unparalleled insights into his development of this momentous modern concept. Bibliography. Index. vii + 121pp. 5⅜ x 8.
S650 Paperbound **$1.15**

A SOURCE BOOK IN MATHEMATICS, D. E. Smith. English translations of the original papers that announced the great discoveries in mathematics from the Renaissance to the end of the 19th century: succinct selections from 125 different treatises and articles, most of them unavailable elsewhere in English—Newton, Leibniz, Pascal, Riemann, Bernoulli, etc. 24 articles trace developments in the field of number, 18 cover algebra, 36 are on geometry, and 13 on calculus. Biographical-historical introductions to each article. Two volume set. Index in each. Total of 115 illustrations. Total of xxviii + 742pp. 5⅜ x 8. S552 Vol I Paperbound **$2.00**
S553 Vol II Paperbound **$2.00**
The set, boxed **$4.00**

CATALOGUE OF DOVER BOOKS

*THE THIRTEEN BOOKS OF EUCLID'S ELEMENTS, edited by T. L. Heath. This is the complete EUCLID — the definitive edition of one of the greatest classics of the western world. Complete English translation of the Heiberg text with spurious Book XIV. Detailed 150-page introduction discusses aspects of Greek and medieval mathematics: Euclid, texts, commentators, etc. Paralleling the text is an elaborate critical exposition analyzing each definition, proposition, postulate, etc., and covering textual matters, mathematical analyses, refutations, extensions, etc. Unabridged reproduction of the Cambridge 2nd edition. 3 volumes. Total of 995. figures, 1426pp. 5⅜ x 8. S88, 89, 90 — 3 vol. set, Paperbound **$6.75**

*THE GEOMETRY OF RENE DESCARTES. The great work which founded analytic geometry. The renowned Smith-Latham translation faced with the original French text containing all of Descartes' own diagrams! Contains: Problems the Construction of Which Requires Only Straight Lines and Circles; On the Nature of Curved Lines; On the Construction of Solid or Supersolid Problems. Notes. Diagrams. 258pp. S68 Paperbound **$1.60**

*A PHILOSOPHICAL ESSAY ON PROBABILITIES, P. Laplace. Without recourse to any mathematics above grammar school, Laplace develops a philosophically, mathematically and historically classical exposition of the nature of probability: its functions and limitations, operations in practical affairs, calculations in games of chance, insurance, government, astronomy, and countless other fields. New introduction by E. T. Bell. viii + 196pp. S166 Paperbound **$1.35**

DE RE METALLICA, Georgius Agricola. Written over 400 years ago, for 200 years the most authoritative first-hand account of the production of metals, translated in 1912 by former President Herbert Hoover and his wife, and today still one of the most beautiful and fascinating volumes ever produced in the history of science! 12 books, exhaustively annotated, give a wonderfully lucid and vivid picture of the history of mining, selection of sites, types of deposits, excavating pits, sinking shafts, ventilating, pumps, crushing machinery, assaying, smelting, refining metals, making salt, alum, nitre, glass, and many other topics. This definitive edition contains all 289 of the 16th century woodcuts which made the original an artistic masterpiece. It makes a superb gift for geologists, engineers, libraries, artists, historians, and everyone interested in science and early illustrative art. Biographical, historical introductions. Bibliography, survey of ancient authors. Indices. 289 illustrations. 672pp. 6¾ x 10¾. Deluxe library edition. S6 Clothbound **$10.00**

GEOGRAPHICAL ESSAYS, W. M. Davis. Modern geography and geomorphology rest on the fundamental work of this scientist. His new concepts of earth-processes revolutionized science and his broad interpretation of the scope of geography created a deeper understanding of the interrelation of the landscape and the forces that mold it. This first inexpensive unabridged edition covers theory of geography, methods of advanced geographic teaching, descriptions of geographic areas, analyses of land-shaping processes, and much besides. Not only a factual and historical classic, it is still widely read for its reflections of modern scientific thought. Introduction. 130 figures. Index. vi + 777pp. 5⅜ x 8. S383 Paperbound **$3.50**

CHARLES BABBAGE AND HIS CALCULATING ENGINES, edited by P. Morrison and E. Morrison. Friend of Darwin, Humboldt, and Laplace, Babbage was a leading pioneer in large-scale mathematical machines and a prophetic herald of modern operational research—true father of Harvard's relay computer Mark I. His Difference Engine and Analytical Engine were the first successful machines in the field. This volume contains a valuable introduction on his life and work; major excerpts from his fascinating autobiography, revealing his eccentric and unusual personality; and extensive selections from "Babbage's Calculating Engines," a compilation of hard-to-find journal articles, both by Babbage and by such eminent contributors as the Countess of Lovelace, L. F. Menabrea, and Dionysius Lardner. 11 illustrations. Appendix of miscellaneous papers. Index. Bibliography. xxxviii + 400pp. 5⅜ x 8. T12 Paperbound **$2.00**

*THE WORKS OF ARCHIMEDES WITH THE METHOD OF ARCHIMEDES, edited by T. L. Heath. All the known works of the greatest mathematician of antiquity including the recently discovered METHOD OF ARCHIMEDES. This last is the only work we have which shows exactly how early mathematicians discovered their proofs before setting them down in their final perfection. A 186 page study by the eminent scholar Heath discusses Archimedes and the history of Greek mathematics. Bibliography. 563pp. 5⅜ x 8. S9 Paperbound **$2.45**

Puzzles, Mathematical Recreations

SYMBOLIC LOGIC and THE GAME OF LOGIC, Lewis Carroll. "Symbolic Logic" is not concerned with modern symbolic logic, but is instead a collection of over 380 problems posed with charm and imagination, using the syllogism, and a fascinating diagrammatic method of drawing conclusions. In "The Game of Logic" Carroll's whimsical imagination devises a logical game played with 2 diagrams and counters (included) to manipulate hundreds of tricky syllogisms. The final section, "Hit or Miss" is a lagniappe of 101 additional puzzles in the delightful Carroll manner. Until this reprint edition, both of these books were rarities costing up to $15 each. Symbolic Logic: Index. xxxi + 199pp. The Game of Logic: 96pp. 2 vols. bound as one. 5⅜ x 8. T492 Paperbound **$1.50**

PILLOW PROBLEMS and A TANGLED TALE, Lewis Carroll. One of the rarest of all Carroll's works, "Pillow Problems" contains 72 original math puzzles, all typically ingenious. Particularly fascinating are Carroll's answers which remain exactly as he thought them out, reflecting his actual mental process. The problems in "A Tangled Tale" are in story form, originally appearing as a monthly magazine serial. Carroll not only gives the solutions, but uses answers sent in by readers to discuss wrong approaches and misleading paths, and grades them for insight. Both of these books were rarities until this edition, "Pillow Problems" costing up to $25, and "A Tangled Tale" $15. Pillow Problems: Preface and Introduction by Lewis Carroll. xx + 109pp. A Tangled Tale: 6 illustrations. 152pp. Two vols. bound as one. 5⅜ x 8. T493 Paperbound **$1.50**

AMUSEMENTS IN MATHEMATICS, Henry Ernest Dudeney. The foremost British originator of mathematical puzzles is always intriguing, witty, and paradoxical in this classic, one of the largest collections of mathematical amusements. More than 430 puzzles, problems, and paradoxes. Mazes and games, problems on number manipulation, unicursal and other route problems, puzzles on measuring, weighing, packing, age, kinship, chessboards, joiners', crossing river, plane figure dissection, and many others. Solutions. More than 450 illustrations. vii + 258pp. 5⅜ x 8. T473 Paperbound **$1.25**

THE CANTERBURY PUZZLES, Henry Dudeney. Chaucer's pilgrims set one another problems in story form. Also Adventures of the Puzzle Club, the Strange Escape of the King's Jester, the Monks of Riddlewell, the Squire's Christmas Puzzle Party, and others. All puzzles are original, based on dissecting plane figures, arithmetic, algebra, elementary calculus and other branches of mathematics, and purely logical ingenuity. "The limit of ingenuity and intricacy," The Observer. Over 110 puzzles. Full Solutions. 150 illustrations. vii + 225pp. 5⅜ x 8. T474 Paperbound **$1.25**

MATHEMATICAL EXCURSIONS, H. A. Merrill. Even if you hardly remember your high school math, you'll enjoy the 90 stimulating problems contained in this book and you will come to understand a great many mathematical principles with surprisingly little effort. Many useful shortcuts and diversions not generally known are included: division by inspection, Russian peasant multiplication, memory systems for pi, building odd and even magic squares, square roots by geometry, dyadic systems, and many more. Solutions to difficult problems. 50 illustrations. 145pp. 5⅜ x 8. T350 Paperbound **$1.00**

MAGIC SQUARES AND CUBES, W. S. Andrews. Only book-length treatment in English, a thorough non-technical description and analysis. Here are nasik, overlapping, pandiagonal, serrated squares; magic circles, cubes, spheres, rhombuses. Try your hand at 4-dimensional magical figures! Much unusual folklore and tradition included. High school algebra is sufficient. 754 diagrams and illustrations. viii + 419pp. 5⅜ x 8. T658 Paperbound **$1.85**

CALIBAN'S PROBLEM BOOK: MATHEMATICAL, INFERENTIAL AND CRYPTOGRAPHIC PUZZLES, H. Phillips (Caliban), S. T. Shovelton, G. S. Marshall. 105 ingenious problems by the greatest living creator of puzzles based on logic and inference. Rigorous, modern, piquant; reflecting their author's unusual personality, these intermediate and advanced puzzles all involve the ability to reason clearly through complex situations; some call for mathematical knowledge, ranging from algebra to number theory. Solutions. xi + 180pp. 5⅜ x 8. T736 Paperbound **$1.25**

MATHEMATICAL PUZZLES FOR BEGINNERS AND ENTHUSIASTS, G. Mott-Smith. 188 mathematical puzzles based on algebra, dissection of plane figures, permutations, and probability, that will test and improve your powers of inference and interpretation. The Odic Force, The Spider's Cousin, Ellipse Drawing, theory and strategy of card and board games like tit-tat-toe, go moku, salvo, and many others. 100 pages of detailed mathematical explanations. Appendix of primes, square roots, etc. 135 illustrations. 2nd revised edition. 248pp. 5⅜ x 8. T198 Paperbound **$1.00**

MATHEMAGIC, MAGIC PUZZLES, AND GAMES WITH NUMBERS, R. V. Heath. More than 60 new puzzles and stunts based on the properties of numbers. Easy techniques for multiplying large numbers mentally, revealing hidden numbers magically, finding the date of any day in any year, and dozens more. Over 30 pages devoted to magic squares, triangles, cubes, circles, etc. Edited by J. S. Meyer. 76 illustrations. 128pp. 5⅜ x 8. T110 Paperbound **$1.00**

CATALOGUE OF DOVER BOOKS

THE BOOK OF MODERN PUZZLES, G. L. Kaufman. A completely new series of puzzles as fascinating as crossword and deduction puzzles but based upon different principles and techniques. Simple 2-minute teasers, word labyrinths, design and pattern puzzles, logic and observation puzzles — over 150 braincrackers. Answers to all problems. 116 illustrations. 192pp. 5⅜ x 8.
.T143 Paperbound **$1.00**

NEW WORD PUZZLES, G. L. Kaufman. 100 ENTIRELY NEW puzzles based on words and their combinations that will delight crossword puzzle, Scrabble and Jotto fans. Chess words, based on the moves of the chess king; design-onyms, symmetrical designs made of synonyms; rhymed double-crostics; syllable sentences; addle letter anagrams; alphagrams; linkograms; and many others all brand new. Full solutions. Space to work problems. 196 figures. vi + 122pp. 5⅜ x 8.
T344 Paperbound **$1.00**

MAZES AND LABYRINTHS: A BOOK OF PUZZLES, W. Shepherd. Mazes, formerly associated with mystery and ritual, are still among the most intriguing of intellectual puzzles. This is a novel and different collection of 50 amusements that embody the principle of the maze: mazes in the classical tradition; 3-dimensional, ribbon, and Möbius-strip mazes; hidden messages; spatial arrangements; etc.—almost all built on amusing story situations. 84 illustrations. Essay on maze psychology. Solutions. xv + 122pp. 5⅜ x 8.
T731 Paperbound **$1.00**

MAGIC TRICKS & CARD TRICKS, W. Jonson. Two books bound as one. 52 tricks with cards, 37 tricks with coins, bills, eggs, smoke, ribbons, slates, etc. Details on presentation, misdirection, and routining will help you master such famous tricks as the Changing Card, Card in the Pocket, Four Aces, Coin Through the Hand, Bill in the Egg, Afghan Bands, and over 75 others. If you follow the lucid exposition and key diagrams carefully, you will finish these two books with an astonishing mastery of magic. 106 figures. 224pp. 5⅜ x 8. T909 Paperbound **$1.00**

PANORAMA OF MAGIC, Milbourne Christopher. A profusely illustrated history of stage magic, a unique selection of prints and engravings from the author's private collection of magic memorabilia, the largest of its kind. Apparatus, stage settings and costumes; ingenious ads distributed by the performers and satiric broadsides passed around in the streets ridiculing pompous showmen; programs; decorative souvenirs. The lively text, by one of America's foremost professional magicians, is full of anecdotes about almost legendary wizards: Dede, the Egyptian; Philadelphia, the wonder-worker; Robert-Houdin, "the father of modern magic;" Harry Houdini; scores more. Altogether a pleasure package for anyone interested in magic, stage setting and design, ethnology, psychology, or simply in unusual people. A Dover original. 295 illustrations; 8 in full color. Index. viii + 216pp. 8⅜ x 11¼.
T774 Paperbound **$2.25**

HOUDINI ON MAGIC, Harry Houdini. One of the greatest magicians of modern times explains his most prized secrets. How locks are picked, with illustrated picks and skeleton keys; how a girl is sawed into twins; how to walk through a brick wall — Houdini's explanations of 44 stage tricks with many diagrams. Also included is a fascinating discussion of great magicians of the past and the story of his fight against fraudulent mediums and spiritualists. Edited by W.B. Gibson and M.N. Young. Bibliography. 155 figures, photos. xv + 280pp. 5⅜ x 8.
T384 Paperbound **$1.35**

MATHEMATICS, MAGIC AND MYSTERY, Martin Gardner. Why do card tricks work? How do magicians perform astonishing mathematical feats? How is stage mind-reading possible? This is the first book length study explaining the application of probability, set theory, theory of numbers, topology, etc., to achieve many startling tricks. Non-technical, accurate, detailed! 115 sections discuss tricks with cards, dice, coins, knots, geometrical vanishing illusions, how a Curry square "demonstrates" that the sum of the parts may be greater than the whole, and dozens of others. No sleight of hand necessary! 135 illustrations. xii + 174pp. 5⅜ x 8.
T335 Paperbound **$1.00**

EASY-TO-DO ENTERTAINMENTS AND DIVERSIONS WITH COINS, CARDS, STRING, PAPER AND MATCHES, R. M. Abraham. Over 300 tricks, games and puzzles will provide young readers with absorbing fun. Sections on card games; paper-folding; tricks with coins, matches and pieces of string; games for the agile; toy-making from common household objects; mathematical recreations; and 50 miscellaneous pastimes. Anyone in charge of groups of youngsters, including hard-pressed parents, and in need of suggestions on how to keep children sensibly amused and quietly content will find this book indispensable. Clear, simple text, copious number of delightful line drawings and illustrative diagrams. Originally titled "Winter Nights Entertainments." Introduction by Lord Baden Powell. 329 illustrations. v + 186pp. 5⅜ x 8½.
T921 Paperbound **$1.00**

STRING FIGURES AND HOW TO MAKE THEM, Caroline Furness Jayne. 107 string figures plus variations selected from the best primitive and modern examples developed by Navajo, Apache, pygmies of Africa, Eskimo, in Europe, Australia, China, etc. The most readily understandable, easy-to-follow book in English on perennially popular recreation. Crystal-clear exposition; step-by-step diagrams. Everyone from kindergarten children to adults looking for unusual diversion will be endlessly amused. Index. Bibliography. Introduction by A. C. Haddon. 17 full-page plates. 960 illustrations. xxiii + 401pp. 5⅜ x 8½.
T152 Paperbound **$2.00**

Americana

THE EYES OF DISCOVERY, J. Bakeless. A vivid reconstruction of how unspoiled America appeared to the first white men. Authentic and enlightening accounts of Hudson's landing in New York, Coronado's trek through the Southwest; scores of explorers, settlers, trappers, soldiers. America's pristine flora, fauna, and Indians in every region and state in fresh and unusual new aspects. "A fascinating view of what the land was like before the first highway went through," Time. 68 contemporary illustrations, 39 newly added in this edition. Index. Bibliography. x + 500pp. 5⅜ x 8.   T761 Paperbound **$2.00**

AUDUBON AND HIS JOURNALS, J. J. Audubon. A collection of fascinating accounts of Europe and America in the early 1800's through Audubon's own eyes. Includes the Missouri River Journals —an eventful trip through America's untouched heartland, the Labrador Journals, the European Journals, the famous "Episodes", and other rare Audubon material, including the descriptive chapters from the original letterpress edition of the "Ornithological Studies", omitted in all later editions. Indispensable for ornithologists, naturalists, and all lovers of Americana and adventure. 70-page biography by Audubon's granddaughter. 38 illustrations. Total of 1106pp. 5⅜ x 8.   T675 Vol I Paperbound **$2.25**
T676 Vol II Paperbound **$2.25**
The set **$4.50**

TRAVELS OF WILLIAM BARTRAM, edited by Mark Van Doren. The first inexpensive illustrated edition of one of the 18th century's most delightful books is an excellent source of first-hand material on American geography, anthropology, and natural history. Many descriptions of early Indian tribes are our only source of information on them prior to the infiltration of the white man. "The mind of a scientist with the soul of a poet," John Livingston Lowes. 13 original illustrations and maps. Edited with an introduction by Mark Van Doren. 448pp. 5⅜ x 8.   T13 Paperbound **$2.00**

GARRETS AND PRETENDERS: A HISTORY OF BOHEMIANISM IN AMERICA, A. Parry. The colorful and fantastic history of American Bohemianism from Poe to Kerouac. This is the only complete record of hoboes, cranks, starving poets, and suicides. Here are Pfaff, Whitman, Crane, Bierce, Pound, and many others. New chapters by the author and by H. T. Moore bring this thorough and well-documented history down to the Beatniks. "An excellent account," N. Y. Times. Scores of cartoons, drawings, and caricatures. Bibliography. Index. xxviii + 421pp. 5⅝ x 8⅜.   T708 Paperbound **$1.95**

THE EXPLORATION OF THE COLORADO RIVER AND ITS CANYONS, J. W. Powell. The thrilling first-hand account of the expedition that filled in the last white space on the map of the United States. Rapids, famine, hostile Indians, and mutiny are among the perils encountered as the unknown Colorado Valley reveals its secrets. This is the only uncut version of Major Powell's classic of exploration that has been printed in the last 60 years. Includes later reflections and subsequent expedition. 250 illustrations, new map. 400pp. 5⅝ x 8⅜.   T94 Paperbound **$2.25**

THE JOURNAL OF HENRY D. THOREAU, Edited by Bradford Torrey and Francis H. Allen. Henry Thoreau is not only one of the most important figures in American literature and social thought; his voluminous journals (from which his books emerged as selections and crystallizations) constitute both the longest, most sensitive record of personal internal development and a most penetrating description of a historical moment in American culture. This present set, which was first issued in fourteen volumes, contains Thoreau's entire journals from 1837 to 1862, with the exception of the lost years which were found only recently. We are reissuing it, complete and unabridged, with a new introduction by Walter Harding, Secretary of the Thoreau Society. Fourteen volumes reissued in two volumes. Foreword by Henry Seidel Canby. Total of 1888pp. 8⅜ x 12¼.   T312-3 Two volume set, Clothbound **$20.00**

GAMES AND SONGS OF AMERICAN CHILDREN, collected by William Wells Newell. A remarkable collection of 190 games with songs that accompany many of them; cross references to show similarities, differences among them; variations; musical notation for 38 songs. Textual discussions show relations with folk-drama and other aspects of folk tradition. Grouped into categories for ready comparative study: Love-games, histories, playing at work, human life, bird and beast, mythology, guessing-games, etc. New introduction covers relations of songs and dances to timeless heritage of folklore, biographical sketch of Newell, other pertinent data. A good source of inspiration for those in charge of groups of children and a valuable reference for anthropologists, sociologists, psychiatrists. Introduction by Carl Withers. New indexes of first lines, games. 5⅜ x 8½. xii + 242pp.   T354 Paperbound **$1.75**

GARDNER'S PHOTOGRAPHIC SKETCH BOOK OF THE CIVIL WAR, Alexander Gardner. The first published collection of Civil War photographs, by one of the two or three most famous photographers of the era, outstandingly reproduced from the original positives. Scenes of crucial battles: Appomattox, Manassas, Mechanicsville, Bull Run, Yorktown, Fredericksburg, etc. Gettysburg immediately after retirement of forces. Battle ruins at Richmond, Petersburg, Gaines'Mill. Prisons, arsenals, a slave pen, fortifications, headquarters, pontoon bridges, soldiers, a field hospital. A unique glimpse into the realities of one of the bloodiest wars in history, with an introductory text to each picture by Gardner himself. Until this edition, there were only five known copies in libraries, and fewer in private hands, one of which sold at auction in 1952 for $425. Introduction by E. F. Bleiler. 100 full page 7 x 10 photographs (original size). 224pp. 8½ x 10¾.   T476 Clothbound **$6.00**

A BIBLIOGRAPHY OF NORTH AMERICAN FOLKLORE AND FOLKSONG, Charles Haywood, Ph.D. The only book that brings together bibliographic information on so wide a range of folklore material. Lists practically everything published about American folksongs, ballads, dances, folk beliefs and practices, popular music, tales, similar material—more than 35,000 titles of books, articles, periodicals, monographs, music publications, phonograph records. Each entry complete with author, title, date and place of publication, arranger and performer of particular examples of folk music, many with Dr. Haywood's valuable criticism, evaluation. Volume I, "The American People," is complete listing of general and regional studies, titles of tales and songs of Negro and non-English speaking groups and where to find them, Occupational Bibliography including sections listing sources of information, folk material on cowboys, riverboat men, 49ers, American characters like Mike Fink, Frankie and Johnnie, John Henry, many more. Volume II, "The American Indian," tells where to find information on dances, myths, songs, ritual of more than 250 tribes in U.S., Canada. A monumental product of 10 years' labor, carefully classified for easy use. "All students of this subject . . . will find themselves in debt to Professor Haywood," Stith Thompson, in American Anthropologist. ". . . a most useful and excellent work," Duncan Emrich, Chief Folklore Section, Library of Congress, in "Notes." Corrected, enlarged republication of 1951 edition. New Preface. New index of composers, arrangers, performers. General index of more than 15,000 items. Two volumes. Total of 1301pp. 6⅛ x 9¼.   T797-798 Clothbound **$12.50**

INCIDENTS OF TRAVEL IN YUCATAN, John L. Stephens. One of first white men to penetrate interior of Yucatan tells the thrilling story of his discoveries of 44 cities, remains of once-powerful Maya civilization. Compelling text combines narrative power with historical significance as it takes you through heat, dust, storms of Yucatan; native festivals with brutal bull fights; great ruined temples atop man-made mounds. Countless idols, sculptures, tombs, examples of Mayan taste for rich ornamentation, from gateways to personal trinkets, accurately illustrated, discussed in text. Will appeal to those interested in ancient civilizations, and those who like stories of exploration, discovery, adventure. Republication of last (1843) edition. 124 illustrations by English artist, F. Catherwood. Appendix on Mayan architecture, chronology. Two volume set. Total of xxviii + 927pp.

Vol I T926 Paperbound **$2.00**
Vol II T927 Paperbound **$2.00**
The set **$4.00**

A GENIUS IN THE FAMILY, Hiram Percy Maxim. Sir Hiram Stevens Maxim was known to the public as the inventive genius who created the Maxim gun, automatic sprinkler, and a heavier-than-air plane that got off the ground in 1894. Here, his son reminisces—this is by no means a formal biography—about the exciting and often downright scandalous private life of his brilliant, eccentric father. A warm and winning portrait of a prankish, mischievous, impious personality, a genuine character. The style is fresh and direct, the effect is unadulterated pleasure. "A book of charm and lasting humor . . . belongs on the 'must read' list of all fathers," New York Times. "A truly gorgeous affair," New Statesman and Nation. 17 illustrations, 16 specially for this edition. viii + 108pp. 5⅜ x 8½.
T948 Paperbound **$1.00**

HORSELESS CARRIAGE DAYS, Hiram P. Maxim. The best account of an important technological revolution by one of its leading figures. The delightful and rewarding story of the author's experiments with the exact combustibility of gasoline, stopping and starting mechanisms, carriage design, and engines. Captures remarkably well the flavor of an age of scoffers and rival inventors not above sabotage; of noisy, uncontrollable gasoline vehicles and incredible mobile steam kettles. ". . . historic information and light humor are combined to furnish highly entertaining reading," New York Times. 56 photographs, 12 specially for this edition. xi + 175pp. 5⅜ x 8½.   T964 Paperbound **$1.35**

BODY, BOOTS AND BRITCHES: FOLKTALES, BALLADS AND SPEECH FROM COUNTRY NEW YORK, Harold W. Thompson. A unique collection, discussion of songs, stories, anecdotes, proverbs handed down orally from Scotch-Irish grandfathers, German nurse-maids, Negro workmen, gathered from all over Upper New York State. Tall tales by and about lumbermen and pirates, canalers and injun-fighters, tragic and comic ballads, scores of sayings and proverbs all tied together by an informative, delightful narrative by former president of New York Historical Society. ". . . a sparkling homespun tapestry that every lover of Americana will want to have around the house," Carl Carmer, New York Times. Republication of 1939 edition. 20 line-drawings. Index. Appendix (Sources of material, bibliography). 530pp. 5⅜ x 8½.   T411 Paperbound **$2.25**

Psychology

YOGA: A SCIENTIFIC EVALUATION, Kovoor T. Behanan. A complete reprinting of the book that for the first time gave Western readers a sane, scientific explanation and analysis of yoga. The author draws on controlled laboratory experiments and personal records of a year as a disciple of a yoga, to investigate yoga psychology, concepts of knowledge, physiology, "supernatural" phenomena, and the ability to tap the deepest human powers. In this study under the auspices of Yale University Institute of Human Relations, the strictest principles of physiological and psychological inquiry are followed throughout. Foreword by W. A. Miles, Yale University. 17 photographs. Glossary. Index. xx + 270pp. 5⅜ x 8. T505 Paperbound **$2.00**

CONDITIONED REFLEXES: AN INVESTIGATION OF THE PHYSIOLOGICAL ACTIVITIES OF THE CEREBRAL CORTEX, I. P. Pavlov. Full, authorized translation of Pavlov's own survey of his work in experimental psychology reviews entire course of experiments, summarizes conclusions, outlines psychological system based on famous "conditioned reflex" concept. Details of technical means used in experiments, observations on formation of conditioned reflexes, function of cerebral hemispheres, results of damage, nature of sleep, typology of nervous system, significance of experiments for human psychology. Trans. by Dr. G. V. Anrep, Cambridge Univ. 235-item bibliography. 18 figures. 445pp. 5⅜ x 8. S614 Paperbound **$2.35**

EXPLANATION OF HUMAN BEHAVIOUR, F. V. Smith. A major intermediate-level introduction to and criticism of 8 complete systems of the psychology of human behavior, with unusual emphasis on theory of investigation and methodology. Part I is an illuminating analysis of the problems involved in the explanation of observed phenomena, and the differing viewpoints on the nature of causality. Parts II and III are a closely detailed survey of the systems of McDougall, Gordon Allport, Lewin, the Gestalt group, Freud, Watson, Hull, and Tolman. Biographical notes. Bibliography of over 800 items. 2 Indexes. 38 figures. xii + 460pp. 5½ x 8¾. T253 Clothbound **$6.00**

SEX IN PSYCHO-ANALYSIS (formerly CONTRIBUTIONS TO PSYCHO-ANALYSIS), S. Ferenczi. Written by an associate of Freud, this volume presents countless insights on such topics as impotence, transference, analysis and children, dreams, symbols, obscene words, masturbation and male homosexuality, paranoia and psycho-analysis, the sense of reality, hypnotism and therapy, and many others. Also includes full text of THE DEVELOPMENT OF PSYCHO-ANALYSIS by Ferenczi and Otto Rank. Two books bound as one. Total of 406pp. 5⅜ x 8. T324 Paperbound **$1.85**

BEYOND PSYCHOLOGY, Otto Rank. One of Rank's most mature contributions, focussing on the irrational basis of human behavior as a basic fact of our lives. The psychoanalytic techniques of myth analysis trace to their source the ultimates of human existence: fear of death, personality, the social organization, the need for love and creativity, etc. Dr. Rank finds them stemming from a common irrational source, man's fear of final destruction. A seminal work in modern psychology, this work sheds light on areas ranging from the concept of immortal soul to the sources of state power. 291pp. 5⅜ x 8. T485 Paperbound **$2.00**

ILLUSIONS AND DELUSIONS OF THE SUPERNATURAL AND THE OCCULT, D. H. Rawcliffe. Holds up to rational examination hundreds of persistent delusions including crystal gazing, automatic writing, table turning, mediumistic trances, mental healing, stigmata, lycanthropy, live burial, the Indian Rope Trick, spiritualism, dowsing, telepathy, clairvoyance, ghosts, ESP, etc. The author explains and exposes the mental and physical deceptions involved, making this not only an exposé of supernatural phenomena, but a valuable exposition of characteristic types of abnormal psychology. Originally titled "The Psychology of the Occult." 14 illustrations. Index. 551pp. 5⅜ x 8. T503 Paperbound **$2.00**

THE PRINCIPLES OF PSYCHOLOGY, William James. The full long-course, unabridged, of one of the great classics of Western literature and science. Wonderfully lucid descriptions of human mental activity, the stream of thought, consciousness, time perception, memory, imagination, emotions, reason, abnormal phenomena, and similar topics. Original contributions are integrated with the work of such men as Berkeley, Binet, Mills, Darwin, Hume, Kant, Royce, Schopenhauer, Spinoza, Locke, Descartes, Galton, Wundt, Lotze, Herbart, Fechner, and scores of others. All contrasting interpretations of mental phenomena are examined in detail — introspective analysis, philosophical interpretation, and experimental research. "A classic," JOURNAL OF CONSULTING PSYCHOLOGY. "The main lines are as valid as ever," PSYCHO-ANALYTICAL QUARTERLY. "Standard reading . . . a classic of interpretation," PSYCHIATRIC QUARTERLY. 94 illustrations. 1408pp. 2 volumes. 5⅜ x 8. Vol. 1, T381 Paperbound **$2.50** Vol. 2, T382 Paperbound **$2.50**

THE DYNAMICS OF THERAPY IN A CONTROLLED RELATIONSHIP, Jessie Taft. One of the most important works in literature of child psychology, out of print for 25 years. Outstanding disciple of Rank describes all aspects of relationship or Rankian therapy through concise, simple elucidation of theory underlying her actual contacts with two seven-year olds. Therapists, social caseworkers, psychologists, counselors, and laymen who work with children will all find this important work an invaluable summation of method, theory of child psychology. xix + 296pp. 5⅜ x 8. T325 Paperbound **$1.75**

CATALOGUE OF DOVER BOOKS

SELECTED PAPERS ON HUMAN FACTORS IN THE DESIGN AND USE OF CONTROL SYSTEMS, Edited by H. Wallace Sinaiko. Nine of the most important papers in this area of increasing interest and rapid growth. All design engineers who have encountered problems involving man as a system-component will find this volume indispensable, both for its detailed information about man's unique capacities and defects, and for its comprehensive bibliography of articles and journals in the human-factors field. Contributors include Chapanis, Birmingham, Adams, Fitts and Jones, etc. on such topics as Theory and Methods for Analyzing Errors in Man-Machine Systems, A Design Philosophy for Man-Machine Control Systems, Man's Senses as Informational Channels, The Measurement of Human Performance, Analysis of Factors Contributing to 460 "Pilot Error" Experiences, etc. Name, subject indexes. Bibliographies of over 400 items. 27 figures. 8 tables. ix + 405pp. 6⅛ x 9¼. S140 Paperbound **$2.75**

THE ANALYSIS OF SENSATIONS, Ernst Mach. Great study of physiology, psychology of perception, shows Mach's ability to see material freshly, his "incorruptible skepticism and independence." (Einstein). Relation of problems of psychological perception to classical physics, supposed dualism of physical and mental, principle of continuity, evolution of senses, will as organic manifestation, scores of experiments, observations in optics, acoustics, music, graphics, etc. New introduction by T. S. Szasz, M. D. 58 illus. 300-item bibliography. Index. 404pp. 5⅜ x 8. S525 Paperbound **$1.75**

PRINCIPLES OF ANIMAL PSYCHOLOGY, N. R. F. Maier and T. C. Schneirla. The definitive treatment of the development of animal behavior and the comparative psychology of all animals. This edition, corrected by the authors and with a supplement containing 5 of their most important subsequent articles, is a "must" for biologists, psychologists, zoologists, and others. First part of book includes analyses and comparisons of the behavior of characteristic types of animal life—from simple multicellular animals through the evolutionary scale to reptiles and birds, tracing the development of complexity in adaptation. Two-thirds of the book covers mammalian life, developing further the principles arrived at in Part I. New preface by the authors. 153 illustrations and tables. Extensive bibliographic material. Revised indices. xvi + 683pp. 5⅜ x 8½. S1120 Paperbound **$3.00** (tentative)

ERROR AND ECCENTRICITY IN HUMAN BELIEF, Joseph Jastrow. From 180 A.D. to the 1930's, the surprising record of human credulity: witchcraft, miracle workings, animal magnetism, mind-reading, astral-chemistry, dowsing, numerology, etc. The stories and exposures of the theosophy of Madame Blavatsky and her followers, the spiritism of Helene Smith, the imposture of Kaspar Hauser, the history of the Ouija board, the puppets of Dr. Luy, and dozens of other hoaxers and cranks, past and present. "As a potpourri of strange beliefs and ideas, it makes excellent reading," New York Times. Formerly titled "Wish and Wisdom, Episodes in the Vagaries of Belief." Unabridged publication. 56 illustrations and photos. 22 full-page plates. Index. xv + 394pp. 5⅜ x 8½. T986 Paperbound **$1.85**

THE PHYSICAL DIMENSIONS OF CONSCIOUSNESS, Edwin G. Boring. By one of the ranking psychologists of this century, a major work which reflected the logical outcome of a progressive trend in psychological theory—a movement away from dualism toward physicalism. Boring, in this book, salvaged the most important work of the structuralists and helped direct the mainstream of American psychology into the neo-behavioristic channels of today. Unabridged republication of original (1933) edition. New preface by the author. Indexes. 17 illustrations. xviii + 251pp. 5⅜ x 8. S1040 Paperbound **$1.75**

BRAIN MECHANISMS AND INTELLIGENCE: A QUANTITATIVE STUDY OF INJURIES TO THE BRAIN, K. S. Lashley. A major contemporary psychologist examines the influence of brain injuries upon the capacity to learn, retentiveness, the formation of the maze habit, etc. Also: the relation of reduced learning ability to sensory and motor defects, the nature of the deterioration following cerebral lesions, comparison of the rat with other forms, and related matters. New introduction by Prof. D. O. Hebb. Bibliography. Index. xxii + 200pp. 5⅜ x 8½. T1038 Paperbound **$1.75**

Prices subject to change without notice.

Dover publishes books on art, music, philosophy, literature, languages, history, social sciences, psychology, handcrafts, orientalia, puzzles and entertainments, chess, pets and gardens, books explaining science, intermediate and higher mathematics, mathematical physics, engineering, biological sciences, earth sciences, classics of science, etc. Write to:

Dept. catrr.
Dover Publications, Inc.
180 Varick Street, N.Y. 14, N.Y.